Lecture Notes in Computer Science 9818

Commenced Publication in 1973
Founding and Former Series Editors:
Gerhard Goos, Juris Hartmanis, and Jan van Leeuwen

Editorial Board

More information about this series at http://www.springer.com/series/7407

Yannick Rondelez · Damien Woods (Eds.)

DNA Computing and Molecular Programming

22nd International Conference, DNA 22
Munich, Germany, September 4–8, 2016
Proceedings

 Springer

Editors
Yannick Rondelez
ESPCI, Laboratoire Gulliver UMR 7083
Paris
France

Damien Woods
Inria
Paris
France

ISSN 0302-9743   ISSN 1611-3349   (electronic)
Lecture Notes in Computer Science
ISBN 978-3-319-43993-8   ISBN 978-3-319-43994-5   (eBook)
DOI 10.1007/978-3-319-43994-5

Library of Congress Control Number: 2016947199

LNCS Sublibrary: SL1 – Theoretical Computer Science and General Issues

Printed on acid-free paper

This Springer imprint is published by Springer Nature
The registered company is Springer International Publishing AG Switzerland

Preface

This volume contains the papers presented at DNA 22: The 22nd International Conference on DNA Computing and Molecular Programming. The conference was held at Ludwig-Maximilians-Universität (LMU) in Munich, Germany, during September 4–8, 2016, and organized under the auspices of the International Society for Nanoscale Science, Computation and Engineering (ISNSCE). A one-day Symposium on RNA-Based Information Processing was held after the main conference on September 9, 2016. The DNA conference series aims to draw together mathematics, computer science, physics, chemistry, biology, and nanotechnology to address the analysis, design, and synthesis of information-based molecular systems.

Papers and presentations were sought in all areas that relate to biomolecular computing, including, but not restricted to: algorithms and models for computation with biomolecular systems; computational processes in vitro and in vivo; molecular switches, gates, devices, and circuits; molecular folding and self-assembly of nanostructures; analysis and theoretical models of laboratory techniques; molecular motors and molecular robotics; studies of fault-tolerance and error correction; software tools for analysis, simulation, and design; synthetic biology and in vitro evolution; applications in engineering, physics, chemistry, biology, and medicine.

Authors who wished to orally present their work were asked to select one of two submission tracks: Track A (full paper) or Track B (one-page abstract with supplementary document). Track B is primarily for authors submitting experimental results who plan to submit to a journal rather than publish in the conference proceedings. We received 55 submissions for oral presentations: 16 submissions in Track A and 39 submissions in Track B. Each submission was reviewed by at least four reviewers. The Program Committee accepted 11 papers in Track A and 18 papers in Track B. This volume contains the papers accepted for Track A. In addition, we received 95 poster submissions for Track C.

We express our sincere appreciation to our invited speakers: Matthew Cook, Monika Heiner, Yan Liu, Pekka Orponen, Rebecca Schulman, and Bernard Yurke and invited tutorial speakers Thomas Ouldridge, Paul W.K. Rothemund, and Nadrian C. Seeman. We especially thank all of the authors who contributed papers to these proceedings, and who presented papers and posters during the conference. Finally, the editors thank the members of the Program Committee and the additional reviewers for their hard work in reviewing the papers and providing constructive comments to the authors, as well as for taking part in enthusiastic post-review discussions.

June 2016

Damien Woods
Yannick Rondelez

Organization

Program Committee for DNA 22

Yannick Rondelez	CNRS/University of Tokyo, Japan (Co-chair)
Damien Woods	California Institute of Technology, USA (Co-chair)
Ebbe Andersen	Aarhus University, Denmark
Robert Brijder	Hasselt University, Belgium
Luca Cardelli	Microsoft Research, Cambridge, UK
Hendrik Dietz	Technical University Munich, Germany
David Doty	University of California Davis, USA
Andrew Ellington	University of Texas at Austin, USA
Andre Estevez-Torres	Université Pierre et Marie Curie Paris, France
Elisa Franco	University of California at Riverside, USA
Anthony Genot	CNRS/University of Tokyo, Japan
Ashwin Gopinath	California Institute of Technology, USA
Natasha Jonoska	University of South Florida, USA
Lila Kari	University of Western Ontario, Canada
Yonggang Ke	Georgia Tech, USA
Maria Kwiatkowska	Oxford University, UK
Tim Liedl	Ludwig-Maximilians-Universität München, Germany
Chenxiang Lin	Yale University, USA
Yan Liu	Arizona State University, USA
Chengde Mao	Purdue University, USA
Satoshi Murata	Tohoku University, Japan
Niall Murphy	Microsoft Research, UK
Matthew Patitz	University of Arkansas, USA
Andrew Phillips	Microsoft Research, UK
Lulu Qian	California Institute of Technology, USA
John Reif	Duke University, USA
Rebecca Schulman	Johns Hopkins University, USA
Robert Schweller	University of Texas Rio Grande Valley, USA
Shinnosuke Seki	University of Electro-Communications, Tokyo, Japan
William Shih	Harvard University, USA
Friedrich C. Simmel	Technische Universität München, Germany
David Soloveichik	University of Texas at Austin, USA
Chris Thachuk	California Institute of Technology, USA
Erik Winfree	California Institute of Technology, USA
Andrew Winslow	Université libre de Bruxelles, Belgium
Peng Yin	Harvard University, USA

Additional Reviewers

David F. Anderson	Nikhil Gopalkrishnan	Andrew Page
Hieu Bui	Rizal Hariadi	Arivazhagan Rajendran
Milan Ceska	Hiroaki Hata	Trent Rogers
Cameron Chalk	Jacob Hendricks	Thomas Schaus
Mingjie Dai	Aleck Johnsen	Jie Song
Frits Dannenberg	Jongmin Kim	Tianqi Song
Masayuki Endo	Akihiko Konagaya	Sungwook Woo
Sudhanshu Garg	Luca Laurenti	Feng Xuan
Manoj Gopalkrishnan	Peter Minary	

Organizing Committee for DNA 22

Tim Liedl	LMU Munich Germany (Co-chair)
Friedrich Simmel	Technical University of Munich Germany (Co-chair)
Andrea Cooke	LMU Munich Germany (Executive Co-chair)
Hendrik Dietz	Technical University of Munich, Germany
Ralf Jungmann	Max Planck Institute for Biochemistry, Munich, Germany

Steering Committee

Natasha Jonoska	University of South Florida, USA (Chair)
Luca Cardelli	Microsoft Research, Cambridge, UK
Anne Condon	University of British Columbia, Canada
Masami Hagiya	University of Tokyo, Japan
Lila Kari	University of Western Ontario, Canada
Satoshi Kobayashi	University of Electro-Communication, Chofu, Japan
Chengde Mao	Purdue University, USA
Satoshi Murata	Tohoku University, Japan
John Reif	Duke University, USA
Grzegorz Rozenberg	University of Leiden, The Netherlands
Nadrian Seeman	New York University, USA
Friedrich Simmel	Technical University of Munich, Germany
Andrew Turberfield	Oxford University, UK
Hao Yan	Arizona State, USA
Erik Winfree	California Institute of Technology, USA

Sponsors

Deutsche Forschungsgemeinschaft (DFG)
Center for NanoScience, LMU
National Science Foundation, USA
Nanosystems Initiative Munich, LMU
GRK2062 Molecular Principles of Synthetic Biology, LMU Research Training Group
Eurofins Genomics GmbH
Science Services GmbH
baseclick GmbH

Invited Speakers

Can Excitonic Quantum Computers be Constructed by DNA Assembly of Chromophore Networks?

Bernard Yurke

Boise State Univeristy, Boise, ID 83725, USA
bernardyurke@boisestate.edu
http://coen.boisestate.edu/faculty-staff2/bio/?id=2

Abstract. Fluorophores have been employed extensively in DNA nanotechnology, principally in donor-acceptor combinations, enabling Föster resonant energy transfer (FRET) to be used for applications, such as monitoring hybridization reactions and monitoring DNA nanomachine functions. FRET is an energy non-conserving process in which a bundle of energy, referred to as a Frenkel exciton, is transferred from the donor fluorophore to the acceptor. The characteristic length scale at which FRET sets in is called the Föster radius and is typically about 5 nm. If the donor and accepter are brought to within less than 2 nm of each other, the energy transfer can occur in an energy conserving manner referred to as coherent FRET. A Frenkel exciton, undergoing coherent FRET exchange among a cluster of chromophores, spreads out over the cluster in a wave-like manner, referred to as a quantum walk. Frenkel excitons also exhibit particle-like aspects and are best viewed as fully quantum mechanical entities. One manifestation of particle-like behavior is that, when two excitons encounter each other, they can experience a two-body interaction that gives rise to quantum mechanical phase shifts. In order for this to happen the chromophores must possess a permanent electric dipole moment and this requires the chromophores to be asymmetric. These two properties of Frenkel excitons – their wave-like behavior and their two-body interaction – are sufficient to enable universal quantum computation. I will describe how these two features can be exploited to implement a complete set of quantum gates for universal quantum computation. Quantum computing, regardless of its embodiment, is a race against decoherence, the process by which the wave-like behavior is destroyed. Chromophores, residing in buffer and attached to DNA, are in an environment highly susceptible to this process. It remains to be seen whether the decoherence rate can be reduced enough to enable Frenkel excitons to perform universal quantum computation by undergoing a many-body quantum walk over a network of chromophores attached to a DNA scaffold.

From One, Many: Programmably Reconfigurable, Multiscale Materials Built with DNA

Rebecca Schulman

Chemical and Biomolecular Engineering and Computer Science,
Johns Hopkins University, Baltimore, USA

Materials within living systems have complex structure that constantly reorganizes in order to continue to function reliably as the environment changes. Commonly, this structure arises because a simple set of components is reorganized by sensors and activating agents into many different forms. For example, tubulin can be organized into cilia, fibrous networks or machines such as the spindle, and the extracellular matrix, an extended matrix composed of a relatively small number of principle protein components, is continually growing and being digested and remodeled in response to interaction with cells within a tissue. The ability to reuse simple components in different materials allows for rapid reorganization and allows material to have structure across many different length scales.

I will describe how we can use DNA to build dynamically reconfigurable materials on the micron to millimeter scales where the responses to inputs can be precisely programmed. The addition of one or combinations of DNA sequences can create large scale changes in the material, and these changes can alter a material's form at length scales ranging from the nanoscale to the millimeter scale. Further, these materials can be continually reorganized in response to series of multiple inputs, suggesting a route to building materials that continue adapt in complex ways over time to their environment.

From Petri Nets to Partial Differential Equations: A Petri Net Perspective on Systems and Synthetic Biology

Monika Heiner

Brandenburg University of Technology Cottbus-Senftenberg,
Senftenberg, Germany
Monika.Heiner@b-tu.de
http://www-dssz.informatik.tu-cottbus.de/DSSZ/Main/
MonikaHeiner

Petri nets offer a graphical & intuitive notation for biochemical reaction networks, such as gene regulatory, signal transduction or metabolic networks. Moreover, they may serve as an umbrella formalism combining different modelling paradigms, where each perspective contributes to a better understanding of the biochemical system under study. In this spirit of BioModel Engineering, we developed over the last two decades our unifying Petri net framework comprising the traditional time-free Petri nets (PN) as well as quantitative, i.e. time-dependent Petri nets such as stochastic Petri nets (SPN), continuous Petri nets (CPN), and hybrid Petri nets (HPN), as well as their coloured counterparts [1].

Coloured Petri nets permit, among others, the convenient and flexible encoding of spatial attributes, and thus the modelling of processes evolving in time and space, which are usually treated as stochastic or deterministic partial differential equations (PDE). In our approach, the discretisation of space on the modelling level, while traditionally the discretisation is left for the PDE integration methods [2].

Our framework is supported by a related Petri net toolkit comprising Snoopy, Charlie and Marcie. It has been applied to a couple of case studies. Those involving spatial aspects include the Brusselator model to explore Turing patterns [3], C. elegans vulval development, stochastic membrane systems composed of active compartments, Ca2+ channels arranged in two-dimensional space, phase variatiation in bacterial colony growth, and Planar Cell Polarity (PCP) signalling in Drosophila wing. Some of them will be sketched in this talk.

References

1. Blätke, M., Heiner, M., Marwan, W.: BioModel Engineering with Petri Nets, chap. 7, pp. 141–193. Elsevier Inc., March 2015
2. Gilbert, D., Heiner, M., Liu, F., Saunders, N.: Colouring space - a coloured framework for spatial modelling in systems biology. In: Colom, J.-M., Desel, J. (Eds.) PETRI NETS 2013, LNCS 7927, pp. 230–249. Springer, Heidelberg (2013)
3. Liu, F., Blätke, M., Heiner, M., Yang, M.: Modelling and simulating reaction-diffusion systems using coloured Petri nets. Comput. Biol. Med. **53**, 297–308 (2014)

Computing Without Random Access Memory: Cyclic Tag Systems for Proofs and Interpretation

Matthew Cook[1,2]

[1]Institute of Neuroinformatics, University of Zürich, Zürich, Switzerland
[2]Institute of Neuroinformatics, ETH Zürich, Zürich, Switzerland

Most simple models of computation that operate on one-dimensional information require some kind of lookup table to be used at each step of their operation. For example, Turing machines look up the next transition according to their state and the symbol they see on the tape. But in some settings, it is not clear how to achieve this random-access capability. Cyclic tag systems are suited to such settings, stepping steadily through a cyclic list rather than requiring random access. Since cyclic tag systems are universal (i.e. capable of simulating a Turing machine), their simplicity makes them an attractive route for proving that other systems are universal as well, and they have been used to prove universality of systems ranging from cellular automata to RNA oritatami. Their extreme simplicity even makes it possible for them to arise naturally in other systems; they have recently been discovered in a cellular automaton's naturally occurring behavior. This talk will give a brief survey of these results.

DNA Nanotechnology: From Structural Design to Functionality

Yan Liu

School of Molecular Sciences and Biodesign Institute, MDB Arizona State
University, Tempe, AZ 85287, USA
Yan_liu@asu.edu
https://biodesign.asu.edu/yan-liu

I will present the most recent work from our research group, which may include thermodynamics and kinetics of DNA tile based self-assembly processes, new designs of wireframe 2D and 3D DNA origami nanostructures, single stranded DNA and RNA origami based on paranemic crossovers and their applications in directed evolution of bivalent aptamers.

Algorithms, Designs and Tools for 3D Wireframe DNA Origami

Pekka Orponen

Department of Computer Science, Aalto University, 00076, Aalto, Finland

Three-dimensional DNA origami designs based on wireframe structures have recently evolved into an interesting alternative to the more established helix-packing designs: several alternative approaches exist [1–4], and functionalizations are beginning to emerge [5]. Wireframe designs are appealing both because they make more efficient use of DNA scaffold than helix-packing approaches, and because they seem to fold with higher yield and remain more stable in low-salt, physiological buffers conditions [2, 4].

Because of the inherent combinatorial complexity of wireframe designs, automation of the design process is a central task already for exploratory reserch, and even more so when aiming to make the methodology robust and generally available. Thus, computerised tools for aiding the process have been developed [2, 4], and lately also numerical modelling and simulation packages such as CanDo and oxDNA have introduced support for them.[1]

The theory underlying wireframe DNA origami design involves quite a number of interesting algorithmic and graph-theoretic ideas and challenges, including several open problems. In this talk, we discuss these underpinnings from the computer science direction, and also survey the current status of the design and modelling tools.

References

1. Shih, W.M., Quispe, J.D., Joyce, G.F.: A 1.7-kilobase single-stranded DNA that folds into a nanoscale octahedron. Nature **427**, 618–621 (2004). doi:10.1038/nature02307
2. Benson, E., Mohammed, A., Gardell, J., Masich, S., Czeizler, E., Orponen, P., Högberg, B.: DNA rendering of polyhedral meshes at the nanoscale. Nature **523**, 441–444 (2015). doi:10.1038/nature14586
3. Zhang, F., Jiang, S., Wu, S., Li, Y., Mao, C., Liu, Y., Yan, H.: Complex wireframe DNA origami nanostructures with multi-arm junction vertices. Nature Nanotechnol. **10**, 779–785 (2016). doi:10.1038/nnano.2015.162
4. Veneziano, R., Ratanalert, S., Zhang, K., Zhang, F., Yan, H., Chiu, W., Bathe, M.: Designer nanoscale DNA assemblies programmed from the top down. Science (2016, in press). doi:10.1126/science.aaf4388
5. Tian, Y., Wang, T., Liu, W., Xin, H.L., Li, H., Ke, Y., Shih, W.M., Gang, O.: Prescribed nanoparticle cluster architectures and low-dimensional arrays built using octahedral DNA origami frames. Nature Nanotechnol. **10**, 637–645 (2015). doi:10.1038/nnano.2015.105

---

[1] http://cando-dna-origami.org/, http://dna.physics.ox.ac.uk/

Invited Tutorial Speakers

Tutorial Abstract: Controlling Structure and Motion in Multiple Dimensions with DNA Information

Nadrian C. Seeman

Department of Chemistry, New York University, New York, NY 10003, USA
ned.seeman@nyu.edu

The essence of Structural DNA Nanotechnology is the combination of branched DNA molecules combined with interactions that can be prescribed by Watson-Crick base pairing. The key goals of the area include the production of objects, lattices and nanomechanical devices made from DNA, as well as controlling the positions of other materials. This approach began by producing structures needing only toplogical control, to generate knots, polyhedral catenanes, Borromean rings and, using L-nucleotides, a Solomon's knot. By the middle 1990s, geometrical control was achieved, leading well-defined objects, often objects acting as tiles for 2D lattices. In the first decade of this century, the development of DNA origami by Paul Rothemund attracted many investigators to DNA nanotechnology, because of the ease of construction and the reliability of obtaining the product from an M13 single-stranded genome and 200–250 'staple strands'. Somewhat later, Peng Yin's use of 'DNA bricks' led to 2D and 3D objects through a completely automated methodology.

Nanorobotics is a key area of application. We have made robust 2-state and 3-state sequence-dependent devices and bipedal walkers. We have constructed a molecular assembly line using a DNA origami layer and three 2-state devices, so that there are eight different states represented by their arrangements. All eight products can be built from this system.

One of the major aims of DNA-based materials research is to construct complex material patterns that can be reproduced. We have built such a system from DNA origami; it has reached 9 generations of exponential growth directly and 24 generations (with no apparent limit) in punctuated steps.

Wenyan Liu's empirical rule states that the best arrays in multidimensional DNA systems result when helix axes span each dimension. We have self-assembled a 2D crystalline origami array by applying this rule. We used the same rule to self-assemble a 3D crystalline array. We initially reported its crystal structure to 4 Å resolution, but rational design of intermolecular contacts has enabled us to improve the crystal resolution to better than 3 Å. We can use crystals with two molecules in the crystallographic repeat to control the color of the crystals. We can change the color of crystals by doing strand displacement of duplex DNA; we can also color the crystals using triplex formation. When tailed in DNA, we can add semiconductors to the crystals, and follow their transitions by crystal color. The use of the crystals to host guests promises an approach to the organization of macromolecules in 3D. Diffraction of the crystals offers a means to ascertain the successful construction of their targets and the characterization of their guests.

This Research was supported by grants EFRI-1332411 and CCF-1526650 from the NSF, MURI W911NF-11-1-0024 from ARO, N000141110729 from ONR, DE-SC0007991 from DOE for DNA synthesis and partial salary support, and grant GBMF3849 from the Gordon and Betty Moore Foundation.

On the Use of DNA Origami to Align Molecular Devices

Ashwin Gopinath[1], Chris Thachuk[1], David Kirkpatrick[2]
and Paul W.K. Rothemund[1]

[1]California Institute of Technology, Pasadena CA 91125, USA
[2]University of British Columbia, Vancouver BC V6T 1Z4, Canada
ashwing@caltech.edu, thachuk@caltech.edu,
kirk@cs.ubc.ca, pwkr@dna.caltech.edu

Over the last decade DNA origami has matured as a modular technique for self-assembling diverse components, from organic molecules to colloidal nanoparticles, into complex nanodevices. A second technique, "DNA origami placement", allows such origami-templated devices to be precisely positioned within microfabricated structures at a resolution of ~ 10 nm in x and y. This allows the integration of point-like or high-symmetry devices with on-chip electronics or optics to create hybrid structures which use self-assembled devices for their novel functional properties, and use microfabricated structures to interrogate the devices or wire them up into larger architectures. However, many devices of interest are highly asymmetric, and both their up-down orientation as well as their in-plane rotational orientation θ must be controlled. Alignment techniques based on mechanical flows, electric fields, and magnetic fields exist, but they typically align all devices in a single coherent orientation and cannot uniquely orient asymmetric devices such as diodes. Here we report extensions of DNA origami placement which allow high fidelity control over both up-down and rotational orientations: 98 % of appropriately-functionalized DNA origami bind to a semiconductor substrate face-up, and over 98 % of appropriately-shaped origami bind within ± 7 degrees of a unique target orientation. To demonstrate orientation-dependent devices, we show that we can control the polarized emission of fluorescent dyes intercalated into DNA origami. Using the same system we show that we can maximize the coupling of fluorophores to a polarized mode of a photonic crystal cavity, and we construct an ultracompact polarimeter which incorporates over 3000 DNA origami devices having both unique and arbitrary orientations.

The Importance of Thermodynamics for Molecular Systems, and the Importance of Molecular Systems for Thermodynamics

Thomas E. Ouldridge

Department of Mathematics, Imperial College London,
180 Queens Gate, London, SW7 2AZ,
t.ouldridge@imperial.ac.uk
http://www.imperial.ac.uk/people/t.ouldridge

Abstract. Improved understanding of molecular systems has only emphasised the sophistication of networks within the cell. Simultaneously, the advance of DNA nanotechnology, a platform within which reactions can be exquisitely controlled, has made the development of artificial architectures a real possibility. Vital to this progress has been a solid foundation in the thermodynamics of molecular systems. In this tutorial, I will set out the fundamental ways in which thermodynamic principles determine what can be achieved with molecular networks, and at what cost. I will then discuss how, in turn, the need to understand molecular systems is driving the development of a new theory of thermodynamics at the microscopic scale.

Keywords: Thermodynamics · Molecular networks · Stochastic thermodynamics

Contents

Full Papers

A Scheme for Molecular Computation of Maximum Likelihood Estimators for Log-Linear Models

Manoj Gopalkrishnan[(✉)]

Tata Institute of Fundamental Research, Mumbai 400 005, India
manoj.gopalkrishnan@gmail.com
http://www.tcs.tifr.res.in/~manoj

Abstract. We propose a novel molecular computing scheme for statistical inference. We focus on the much-studied statistical inference problem of computing maximum likelihood estimators for log-linear models. Our scheme takes log-linear models to reaction systems, and the observed data to initial conditions, so that the corresponding equilibrium of each reaction system encodes the corresponding maximum likelihood estimator. The main idea is to exploit the coincidence between thermodynamic entropy and statistical entropy. We map a Maximum Entropy characterization of the maximum likelihood estimator onto a Maximum Entropy characterization of the equilibrium concentrations for the reaction system. This allows for an efficient encoding of the problem, and reveals that reaction networks are superbly suited to statistical inference tasks. Such a scheme may also provide a template to understanding how *in vivo* biochemical signaling pathways integrate extensive information about their environment and history.

1 Introduction

The sophisticated behavior of cells emerges from the computations that are being performed by the underlying biochemical reaction networks. These biochemical pathways have been studied in a "top-down" manner, by looking for recurring motifs, and signs of modularity [18]. There is also an opportunity to study these pathways in a "bottom-up" manner by proposing primitive building blocks which can be composed to create interesting and technologically valuable behavior. This "bottom-up" approach connects with work in the Molecular Computation community whose goal is to generate sophisticated behavior using DNA hybridization reactions [3,6,7,19,22–25,27,31] and other Artificial Chemistry approaches [5,10].

We propose a new building block for molecular computation. We show that the mathematical structure of reaction networks is particularly well adapted to compute Maximum Likelihood Estimators for log-linear models, allowing a pithy encoding of such computations by reactions. According to [12]:

© Springer International Publishing Switzerland 2016
Y. Rondelez and D. Woods (Eds.): DNA 2016, LNCS 9818, pp. 3–18, 2016.
DOI: 10.1007/978-3-319-43994-5_1

Log-linear models are arguably the most popular and important statistical models for the analysis of categorical data; see, for example, Bishop, Fienberg and Holland (1975) [4], Christensen (1997) [8]. These powerful models, which include as special cases graphical models [see, e.g., Lauritzen (1996) [16]] as well as many logit models [see, e.g., Agresti (2002) [1], Bishop, Fienberg and Holland (1975) [4]], have applications in many scientific areas, ranging from social and biological sciences, to privacy and disclosure limitation problems, medicine, data mining, language processing and genetics. Their popularity has greatly increased in the last decades...

In order to respond in a manner that maximizes fitness, a cell has to correctly estimate the overall state of its environment. Receptors that sit on cell walls collect a large amount of information about the cellular environment. Processing and integration of this spatially and temporally extensive and diverse information is carried out in the biochemical reaction pathways. We propose that this processing and integration may be advantageously viewed from the lens of machine learning.

Our proposal entails that *schemes for statistical inference by reaction networks are of biological significance, and are deserving of as thorough and extensive a study as schemes for statistical inference by neural networks.* In particular, machine learning is not just a tool for the analysis of biochemical data, but theoretical and technological insights from machine learning could provide a deep and fundamental way, and perhaps "the" correct way, to think about biochemical networks. We view the scheme we present here as a promising first step in this program of applying machine learning insights to biochemical networks.

The problem: We illustrate the main ideas of our scheme with an example. Following [21], consider the **log-linear model** (also known as toric model) described by the **design matrix** $A = \left(\begin{smallmatrix} 2 & 1 & 0 \\ 0 & 1 & 2 \end{smallmatrix} \right)$. This means that we are observing an event with three possible mutually exclusive outcomes, call them X_1, X_2, and X_3, which represent respectively the columns of A. The rows of A represent "hidden variables" θ_1 and θ_2 respectively which parametrize the statistics of the outcomes in the following way specified by the columns of A:

$$P[X_1 \mid \theta_1, \theta_2] \propto \theta_1^2$$
$$P[X_2 \mid \theta_1, \theta_2] \propto \theta_1 \theta_2$$
$$P[X_3 \mid \theta_1, \theta_2] \propto \theta_2^2$$

where the constant of proportionality normalizes the probabilities so they sum to 1.[1]

Suppose several independent trials are carried out, and the outcome X_1 is observed $x_1 \in (0, 1)$ fraction of the time, the outcome X_2 is observed $x_2 \in (0, 1 - x_1)$ fraction of the time, and the outcome X_3 is observed $x_3 = 1 - x_1 - x_2$ fraction

---

[1] It is more common in statistics and statistical mechanics literature to write $\theta_1 = e^{-E_1}$ and $\theta_2 = e^{-E_2}$ in terms of "energies" E_1, E_2 so that $P[X_2 \mid E_1, E_2] \propto e^{-E_1 - E_2}$ for example.

of the time. We wish to find the maximum likelihood estimator $(\hat{\theta}_1, \hat{\theta}_2) \in \mathbb{R}^2_{>0}$ of the parameter (θ_1, θ_2), i.e., that value of θ which maximizes the likelihood of the observed data.

Our contribution: We describe a scheme that takes the design matrix A to a reaction network that solves the maximum likelihood estimation problem. In Definition 8, we describe our scheme for every matrix A over the integers with all column sums equal. All our results hold in this generality.

- In Definition 8. Theorem 5, we show how to obtain from the matrix A, a reaction network that computes the maximum likelihood distribution. Specialized to our example, note that the kernel of the matrix A is spanned by the vector $(1, -2, 1)^T$. We encode this by the reversible reaction

$$X_1 + X_3 \underset{1}{\overset{1}{\rightleftharpoons}} 2X_2$$

- In Theorem 5, we show that if this reversible reaction is started at initial concentrations $X_1(0) = x_1, X_2(0) = x_2, X_3(0) = x_3$, and the dynamics proceeds according to the law of mass action with all specific rates set to 1:

$$\dot{X}_1(t) = \dot{X}_3(t) = -X_1(t)X_3(t) + X_2^2(t), \quad \dot{X}_2(t) = -2X_2^2(t) + 2X_1(t)X_3(t)$$

then the reaction reaches equilibrium $(\hat{x}_1, \hat{x}_2, \hat{x}_3)$ where $\hat{x}_1 + \hat{x}_2 + \hat{x}_3 = 1$ and $\hat{x}_1 \propto \hat{\theta}_1^2$, $\hat{x}_2 \propto \hat{\theta}_1\hat{\theta}_2$, and $\hat{x}_3 \propto \hat{\theta}_2^2$, so that $(\hat{x}_1, \hat{x}_2, \hat{x}_3)$ represents the probability distribution over the outcomes X_1, X_2, X_3 at the maximum likelihood $\hat{\theta}_1, \hat{\theta}_2$.
- This part of our scheme involves only reversible reactions, and requires no catalysis (see [13, Theorem 5.2] and Lemma 2). One difficulty with implementing such schemes has been that empirical control over kinetics is rather poor. Exquisitely setting the specific rates of individual reactions to desired values is very tricky, and requires a detailed understanding of molecular dynamics. Our scheme avoids this problem since any choice of specific rates that leads to the same equilibrium will do. Hence we can freely set the specific rates so long as the equilibrium constants (ratio of forward and backward specific rates) have value 1. This is an equilibrium thermodynamic condition that is much easier to ensure in vitro. This combination of reversible reactions, no catalysis, and robustness to the values of the specific rates may make this scheme particularly easy and efficient to implement.
- In Definition 8.2, we show how to obtain from the matrix A a reaction network that computes the maximum likelihood estimator. Specialized to our example, we obtain the reaction network with 5 species $X_1, X_2, X_3, \theta_1, \theta_2$ and the 5 reactions:

$$X_1 + X_3 \rightleftharpoons 2X_2, \qquad 2\theta_1 \rightarrow 0, \qquad X_1 \rightarrow X_1 + 2\theta_1,$$
$$\theta_1 + \theta_2 \rightarrow 0, \qquad X_2 \rightarrow X_2 + \theta_1 + \theta_2.$$

The number of species equals the number of rows plus the number of columns of A. The reactions are not uniquely determined by the problem, but become so once we choose a basis for the kernel of A and a maximal linearly independent set of columns. Here we have chosen columns 1 and 2. Each column of A determines a pair of irreversible reactions.

– Theorem 6 implies that if this reaction system is launched at initial concentrations $X_1(0) = x_1, X_2(0) = x_2, X_3(0) = x_3$ and arbitrary concentrations of $\theta_1(0)$ and $\theta_2(0)$, and the dynamics proceeds according to the law of mass action with all specific rates set to 1:

$$\dot{X}_1(t) = \dot{X}_3(t) = -X_1(t)X_3(t) + X_2^2(t), \dot{X}_2(t) = -2X_2^2(t) + 2X_1(t)X_3(t),$$
$$\dot{\theta}_1(t) = -2\theta_1^2(t) + 2X_1(t) - \theta_1(t)\theta_2(t) + X_2(t), \dot{\theta}_2(t) = -\theta_1\theta_2(t) + X_2(t),$$

then the reaction reaches equilibrium $(\hat{x}_1, \hat{x}_2, \hat{x}_3, \hat{\theta}_1, \hat{\theta}_2)$ where $(\hat{\theta}_1, \hat{\theta}_2)$ is the maximum likelihood estimator for the data frequency vector (x_1, x_2, x_3) and $(\hat{x}_1, \hat{x}_2, \hat{x}_3)$ represents the probability distribution over the outcomes X_1, X_2, X_3 at the maximum likelihood. We prove global convergence: our dynamical system provably converges to the desired equilibrium. Global convergence results are known to be notoriously hard to prove in reaction network theory [14].

– A number of schemes have been proposed for translating reaction networks into DNA strand displacement reactions [6,7,22,27]. Adapting these schemes to our setting should allow molecular implementation of our MLE-solving reaction networks with DNA molecules.

2   Maximum Likelihood Estimation in Toric Models

The definitions and results in this section mostly follow [21]. Because we require a slightly stronger statement, and Theorem 1 allows a short, easy, and insightful proof, we give the proof here for completeness.

In statistics, a **parametric model** consists of a family of probability distributions, one for each value of the parameters. This can be described as a map from a manifold of parameters into a manifold of probability distributions. If this map can be described by monomials as below, then the parametric statistical model is called a **toric** or **log-linear** model, as we now describe.

Definition 1 (Toric Model). *Let m, n be positive integers. The probability simplex and its relative interior are:*

$$\Delta^n := \{(x_1, x_2, \ldots, x_n) \in \mathbb{R}_{\geq 0}^n \mid x_1 + x_2 + \cdots + x_n = 1\}$$

$$\mathrm{ri}(\Delta^n) := \{(x_1, x_2, \ldots, x_n) \in \mathbb{R}_{>0}^n \mid x_1 + x_2 + \cdots + x_n = 1\}.$$

*An $m \times n$ matrix $A = (a_{ij})_{m \times n}$ of integer entries is a **design matrix** iff all its column sums $\sum_i a_{ij}$ are equal. Let $a_j := (a_{1j}, a_{2j}, \ldots, a_{mj})^T$ be the j'th column of A. Define $\theta^{a_j} := \theta_1^{a_{1j}} \theta_2^{a_{2j}} \ldots \theta_m^{a_{mj}}$. Define the **parameter space** $\Theta := \{\theta \in \mathbb{R}_{>0}^m \mid \theta^{a_1} + \theta^{a_2} + \cdots + \theta^{a_n} = 1\}$. The **toric model** of A is the map*

$$p_A = (p_1, p_2, \ldots, p_n) : \Theta \to \Delta^n \text{ given by } p_j(\theta) = \theta^{a_j} \text{ for } j = 1 \text{ to } n.$$

We could also have defined the parameter space Θ to be all of $\mathbb{R}^m_{>0}$, in which case we would need to normalize the probabilities by the *partition function* $\theta^{a_1} + \theta^{a_2} + \cdots + \theta^{a_n}$ to make sure they add up to 1. For our present purposes, the current approach will prove technically more direct.

Note that here $p_j(\theta)$ specifies $\Pr[j \mid \theta]$, the conditional probability of obtaining outcome j given that the true state of the world is described by θ.

A central problem of statistical inference is the problem of **parameter estimation**. After performing several independent identical trials, suppose the **data vector** $u \in \mathbb{Z}^n_{\geq 0}$ is obtained as a record of how many times each outcome occurred. Let the norm $|u|_1 := u_1 + u_2 + \cdots + u_n$ denote the total number of trials performed. The **Maximum Likelihood** solution to the problem of parameter estimation finds that value of the parameter θ which maximizes the **likelihood function** $f_u(\theta) := \Pr[u \mid \theta]$, i.e.:

$$\hat{\theta}(u) := \arg\sup_{\theta \in \Theta} f_u(\theta) \tag{1}$$

is a **maximum likelihood estimator** or MLE for the data vector u. We will call the point $\hat{p}(u) := p_A(\hat{\theta}(u))$ a **maximum likelihood distribution**.

Definition 2. *Let A be an $m \times n$ design matrix, and u a data vector. Then the **sufficient polytope** is $P_A(u) := \{p \in \mathrm{ri}(\Delta^n) \mid Ap = A\frac{u}{|u|_1}\}$.*

The following theorem is a version of Birch's theorem from Algebraic Statistics. It provides a variational characterization of the maximum likelihood distribution as the unique maximum entropy distribution in the sufficient polytope. In particular the maximum likelihood distribution always belongs to the sufficient polytope, which justifies the name.

Theorem 1. *Fix a design matrix A of size $m \times n$.*

1. *If $u, v \in \mathbb{Z}^n_{\geq 0}$ are nonzero data vectors such that $Au/|u|_1 = Av/|v|_1$ then they have the same maximum likelihood estimator: $\hat{\theta}(u) = \hat{\theta}(v)$.*
2. *Further if $P_A(u)$ is nonempty then*
   (a) There is a unique distribution $\tilde{p} \in P_A(u)$ which maximizes Shannon entropy $H(p) = -\sum_{i=1}^n p_i \log p_i$ viewed as a real-valued function from the closure $\overline{P_A(u)}$ of $P_A(u)$ with $0 \log 0$ defined as 0.
   (b) $\{\tilde{p}\} = P_A(u) \cap p_A(\Theta)$.
   (c) $\tilde{p} = \hat{p}(u)$, the Maximum Likelihood Distribution for the data vector u.

Proof. 1. Fix a data vector u. Note that

$$f_u(\theta) = \frac{|u|_1!}{u_1! u_2! \dots u_n!} p_1(\theta)^{u_1} p_2(\theta)^{u_2} \dots p_n(\theta)^{u_n} = \frac{|u|_1!}{u_1! u_2! \dots u_n!} \theta^{Au}.$$

Therefore the maximum likelihood estimator

$$\hat{\theta}(u) = \arg\sup_{\theta \in \Theta} \theta^{Au} = \arg\sup_{\theta \in \Theta} (\theta^{Au})^{1/|u|_1} = \arg\sup_{\theta \in \Theta} \theta^{Au/|u|_1}$$

where the second equality is true because the function $x \mapsto x^c$ is monotonically increasing whenever $c > 0$. It follows that if $v \in \mathbb{Z}_{\geq 0}^n$ is a data vector such that $Au/|u|_1 = Av/|v|_1$ then $\hat{\theta}(u) = \hat{\theta}(v)$.

2. (a) Suppose $P_A(u)$ is nonempty. A local maximum of the restriction $H|_{\overline{P_A(u)}}$ of H to the polytope $\overline{P_A(u)}$ can not be on the boundary $\partial \overline{P_A(u)}$ because for $p \in \partial \overline{P_A(u)}$, moving in the direction of arbitrary $q \in P_A(u)$ increases H, as can be shown by a simple calculation:

$$\lim_{\lambda \to 0} \frac{d}{d\lambda} H((1 - \lambda)p + \lambda q) \to +\infty.$$

Since H is a continuous function and the closure $\overline{P_A(u)}$ is a compact set, H must attain its maximum value in $P_A(u)$. Further H is a strictly concave function since its Hessian is diagonal with entries $-1/p_i$ and hence negative definite. It follows that $H|_{\overline{P_A(u)}}$ is also strictly concave, and has a unique local maximum at $\tilde{p} \in P_A(u)$, which is also the global maximum.

(b) By concavity of H, the maximum $\tilde{p}$ is the unique point in $P_A(u)$ such that $\nabla H(\tilde{p})$ is perpendicular to $P_A(u)$. We claim that $q \in P_A(u) \cap p_A(\Theta)$ iff $\nabla H(q) = (-1 - \log q_1, -1 - \log q_2, \ldots, -1 - \log q_n)$ is perpendicular to $P_A(u)$. Since all column sums are equal, this is equivalent to requiring that $\log q$ be in the span of the rows of A, which is true iff $q \in p_A(\Theta)$. Hence $P_A(u) \cap p_A(\Theta) = \{\tilde{p}\}$.

(c) To compute the Maximum Likelihood Distribution $\hat{p}(u)$, we proceed as follows:

$$\hat{p}(u) = p_A(\hat{\theta}(u)) = p_A(\arg\sup_{\theta \in \Theta} \theta^{Au}) = p_A(\arg\sup_{\theta \in \Theta} \theta^{Au/|u|_1})$$

$$= p_A(\arg\sup_{\theta \in \Theta} \theta^{A\tilde{p}}) = \arg\sup_{p \in p_A(\Theta)} p^{\tilde{p}} = \arg\sup_{p \in p_A(\Theta)} \sum_{i=1}^n \tilde{p}_i \log p_i = \tilde{p}$$

where the fourth equality uses $A\tilde{p} = Au/|u|_1$ and the last equality follows because $\sum_{i=1}^n \tilde{p}_i \log p_i$ viewed as a function of p attains its maximum in all of Δ^n, and hence in $p_A(\Theta)$, at $p = \tilde{p}$.

This theorem already exposes the core of our idea. We will design reaction systems that maximize entropy subject to the "correct" constraints capturing the polytope $P_A(u)$. Then because the reactions also proceed to maximize entropy, the equilibrium point of our dynamics will correspond to the maximum likelihood distribution. Most of the technical work will go in proving convergence of trajectories to these equilibrium points.

3 Reaction Networks

According to [20], "In building a design theory for chemistry, chemical reaction networks are usually the most natural intermediate representation - the middle of the hourglass [11]. Many different high level languages and formalisms have been and can likely be compiled to chemical reactions, and chemical reactions themselves (as an abstract specification) can be implemented with a variety of low level molecular mechanisms."

In Subsect. 3.1, we recall the definitions and results for reaction networks which we will need for our main results. For a comprehensive presentation of these ideas, see [13]. In Subsect. 3.2, we prove a new result in reaction network theory. We extend a previously known global convergence result to the case of perturbations.

3.1   Brief Review of Reaction Network Theory

For vectors $a = (a_i)_{i \in S}$ and $b = (b_i)_{i \in S}$, the notation a^b will be shorthand for the formal monomial $\prod_{i \in S} a_i^{b_i}$. We introduce some standard definitions.

Definition 3 (Reaction Network). *Fix a finite set S of **species**.*

1. *A **reaction** over S is a pair (y, y') such that $y, y' \in \mathbb{Z}_{\geq 0}^S$. It is usually written $y \to y'$, with **reactant** y and **product** y'.*
2. *A **reaction network** consists of a finite set S of species, and a finite set $\mathcal{R}$ of reactions.*
3. *A reaction network is **reversible** iff for every reaction $y \to y' \in \mathcal{R}$, the reaction $y' \to y \in \mathcal{R}$.*
4. *A reaction network is **weakly reversible** iff for every reaction $y \to y' \in \mathcal{R}$ there exists a positive integer $n \in \mathbb{Z}_{>0}$ and n reactions $y_1 \to y_2, y_2 \to y_3, \ldots, y_{n-1} \to y_n \in \mathcal{R}$ with $y_1 = y'$ and $y_n = y$.*
5. *The **stoichiometric subspace** $H \subseteq \mathbb{R}^S$ is the subspace spanned by $\{y' - y \mid y \to y' \in \mathcal{R}\}$, and $H^\perp$ is the orthogonal complement of H.*
6. *A **siphon** is a set $T \subseteq S$ of species such that for all $y \to y' \in \mathcal{R}$, if there exists $i \in T$ such that $y'_i > 0$ then there exists $j \in T$ such that $y_j > 0$.*
7. *A siphon $T \subseteq S$ is **critical** iff $v \in H^\perp \cap \mathbb{R}_{\geq 0}^S$ with $v_i = 0$ for all $i \notin T$ implies $v = 0$.*

Definition 4. *Fix a weakly reversible reaction network $(S, \mathcal{R})$. The **associated ideal** $I_{(S,\mathcal{R})} \subseteq \mathbb{C}[x]$ where $x = (x_i)_{i \in S}$ is the ideal generated by the binomials $\{x^y - x^{y'} \mid y \to y' \in \mathcal{R}\}$. A reaction network is **prime** iff its associated ideal is a prime ideal.*

The following theorem follows from [13, Theorems 4.1 and 5.2].

Theorem 2. *A weakly reversible prime reaction network $(S, \mathcal{R})$ has no critical siphons.*

We now recall the mass-action equations which are widely employed for modeling cellular processes [26, 28–30] in Biology.

Definition 5 (Mass Action System). *A **reaction system** consists of a reaction network $(S, \mathcal{R})$ and a **rate function** $k : \mathcal{R} \to \mathbb{R}_{>0}$. The **mass-action**

equations for a reaction system are the system of ordinary differential equations in concentration *variables* $\{x_i(t) \mid i \in S\}$:

$$\dot{x}(t) = \sum_{y \to y' \in \mathbb{R}} k_{y \to y'}\, x(t)^y\, (y' - y) \tag{2}$$

where $x(t)$ *represents the vector* $(x_i(t))_{i \in S}$ *of concentrations at time* t.

Note that $\dot{x}(t) \in H$, so affine translations of H are invariant under the dynamics of Eq. 2.

We recall the well known notions of detailed balanced and complex balanced reaction system.

Definition 6. *A reaction system* $(S, \mathcal{R}, k)$ *is*

1. **Detailed balanced** *iff it is reversible and there exists a point* $\alpha \in \mathbb{R}^S_{>0}$ *such that for every* $y \to y' \in \mathcal{R}$:

$$k_{y \to y'}\, \alpha^y\, (y' - y) = k_{y' \to y}\, \alpha^{y'}\, (y - y')$$

A point $\alpha \in \mathbb{R}^S_{>0}$ *that satisfies the above condition is called a* **point of detailed balance**.

2. **Complex balanced** *iff there exists a point* $\alpha \in \mathbb{R}^S_{>0}$ *such that for every* $y \in \mathbb{Z}^S_{\geq 0}$:

$$\sum_{y \to y' \in \mathcal{R}} k_{y \to y'}\, \alpha^y\, (y' - y) = \sum_{y'' \to y \in \mathcal{R}} k_{y'' \to y}\, \alpha^{y''}\, (y - y'')$$

A point $\alpha \in \mathbb{R}^S_{>0}$ *that satisfies the above condition is called a* **point of complex balance**.

The following observations are well known and easy to verify.

- A complex balanced reaction system is always weakly reversible.
- If all rates $k_{y \to y'} = 1$ and the network is weakly reversible then the reaction system is complex balanced with point of complex balance $(1, 1, \ldots, 1) \in \mathbb{R}^S$; if the network is reversible then the reaction system is also detailed balanced with point of detailed balance $(1, 1, \ldots, 1) \in \mathbb{R}^S$.
- Every detailed balance point is also a complex balance point, but there are complex balanced reversible networks that are not detailed balanced.

It is straightforward to check that every point of complex balance (respectively, detailed balance) is a fixed point for Eq. 2. The next theorem, which follows from [2, Theorem 2] and [15], states that a converse also exists: if a reaction system is complex balanced (respectively, detailed balanced) then every fixed point is a point of complex balance (detailed balance). Further there is a unique fixed point in each affine translation of H, and if there are no critical siphons then the basin of attraction for this fixed point is as large as possible, namely the intersection of the affine translation of H with the nonnegative orthant.

Theorem 3 (Global Attractor Theorem for Complex Balanced Reaction Systems with no Critical Siphons). *Let $(S, \mathcal{R}, k)$ be a weakly reversible complex balanced reaction system with no critical siphons and point of complex balance α. Fix a point $u \in \mathbb{R}_{>0}^S$. Then there exists a point of complex balance β in $(u + H) \cap \mathbb{R}_{>0}^S$ such that for every trajectory $x(t)$ with initial conditions $x(0) \in (u + H) \cap \mathbb{R}_{\geq 0}^S$, the limit $\lim_{t \to \infty} x(t)$ exists and equals β. Further the function $g(x) := \sum_{i=1}^{n} x_i \log x_i - x_i - x_i \log \alpha_i$ is strictly decreasing along nonstationary trajectories and attains its unique minimum value in $(u + H) \cap \mathbb{R}_{\geq 0}^S$ at β.*

It is not completely trivial to show, but nevertheless true, that this theorem holds with weakly reversible replaced by "reversible" and "complex balance" replaced by "detailed balance." What is to be shown is that the point of complex balance obtained in $(u + H) \cap \mathbb{R}_{\geq 0}^S$ by minimizing $g(x)$ is actually a point of detailed balance, and this follows from an examination of the form of the derivative $\frac{d}{dt} g(x(t))$ along trajectories $x(t)$ to Eq. 2.

3.2   A Perturbatively-Stable Global Attractor Theorem

Global attractor results usually assume that the reaction network is weakly reversible. We are going to describe our scheme in the next section. Our scheme will employ reaction networks that are not weakly reversible, yet we will prove global attractor results for them. The key idea we use is that our reaction network can be broke into a reversible part, and an irreversible part. The reversible part acts on, but evolves independent of, the irreversible part. So we get to use the global attractor results "as is" on the reversible part. Further, as the reversible part approaches equilibrium, our irreversible part behaves as a perturbation of a reversible detailed-balanced network. The closer the reversible part gets to equilibrium, the smaller the perturbation of the irreversible part from the dynamics of a certain reversible detailed-balanced network.

To make this proof idea work out, we will need a perturbative version of Theorem 3. The next lemma shows that if the rates are perturbed slightly then, outside a small neighborhood of the detailed balance point, the strict Lyapunov function $g(x)$ from Theorem 3 continues to decrease along non-stationary trajectories.

Lemma 1. *Let $(S, \mathcal{R}, k)$ be a weakly reversible complex balanced reaction system with no critical siphons and point of complex balance α. For every sufficiently small $\epsilon > 0$ there exists $\delta > 0$ such that for all x' outside the ϵ-neighborhood of α in $(\alpha + H) \cap \mathbb{R}_{\geq 0}^S$, the derivative $\frac{d}{dt} g(x(t))|_{t=0} < -\delta$, where $x(t)$ is a solution to the Mass-Action Equations 2 with $x(0) = x'$.*

Proof. Let B_ϵ be the open ϵ ball around α in $(\alpha + H) \cap \mathbb{R}_{\geq 0}^S$, with ϵ small enough so that B_ϵ does not meet the boundary $\partial \mathbb{R}_{\geq 0}^S$. Consider the closed set $S := (\alpha + H) \cap \mathbb{R}_{\geq 0}^S \setminus B_\epsilon$. Define the orbital derivative of g at x' as $\mathcal{O}_k g(x') := \frac{d}{dt} g(x(t))|_{t=0}$, where $x(t)$ is a solution to the mass-action equations 2 with $x(0) = x'$.

Define $\delta := \inf_{x' \in S}(-\mathcal{O}_k g(x'))$. If $\delta \leq 0$ then since S is a closed set, and $\mathcal{O}_k g$ is a continuous function, there exists a point x' such that $\mathcal{O}_k g(x') \geq 0$, which contradicts Theorem 3.

We formalize the notion of perturbation using **differential inclusions**. Recall that differential inclusions model uncertainty in dynamics in a nondeterministic way by generalizing the notion of vector field. A differential inclusion maps every point to a subset of the tangent space at that point.

Definition 7. *Let $(S, \mathcal{R}, k)$ be a reaction system and let $\delta > 0$. Then the δ-**perturbation** of $(S, \mathcal{R}, k)$ is the differential inclusion $V : \mathbb{R}_{\geq 0}^S \to 2^{\mathbb{R}^S}$ that at point $x \in \mathbb{R}_{\geq 0}^S$ takes the value $V(x) :=*

$$\left\{ \sum_{y \to y' \in \mathcal{R}} k'_{y \to y'} x^y (y' - y) \,\middle|\, k'_{y \to y'} \in (k_{y \to y'} - \delta, k_{y \to y'} + \delta) \text{ for all } y \to y' \in \mathcal{R} \right\}.$$

*A **trajectory** of V is a tuple (I, x) where $I \subseteq \mathbb{R}$ is an interval and $x : I \to \mathbb{R}_{\geq 0}^S$ is a differentiable function with $\dot{x}(t) \in V(x(t))$.*

Theorem 4 (Perturbatively-Stable Global Attractor Theorem for Complex Balanced Reaction Systems with no Critical Siphons).

Let $(S, \mathcal{R}, k)$ be a weakly reversible complex balanced reaction system with no critical siphons. Fix a point $u \in \mathbb{R}_{>0}^S$. Then there exists a point of complex balance β in $(u + H) \cap \mathbb{R}_{>0}^S$ such that:

1. *For every sufficiently small $\varepsilon > 0$, there exists $\delta > 0$ such that every trajectory of the form $(\mathbb{R}_{\geq 0}, x)$ to the δ-perturbation of $(S, \mathcal{R}, k)$ with initial conditions $x(0) \in (u + H) \cap \mathbb{R}_{\geq 0}^S$ eventually enters an ε-neighborhood of β and never leaves.*
2. *Consider a sequence $\delta_1 > \delta_2 > \cdots > 0$ and a sequence $0 < t_1 < t_2 < \ldots$ such that $\lim_{i \to \infty} \delta_i = 0$ and $\lim_{i \to \infty} t_i = +\infty$, and a trajectory $(\mathbb{R}_{\geq 0}, x)$ with $x(0) \in (u+H) \cap \mathbb{R}_{\geq 0}^S$ such that $((t_i, \infty), x)$ is a trajectory of the δ_i-perturbation of $(S, \mathcal{R}, k)$. Then the limit $\lim_{t \to \infty} x(t) = \beta$.*

Proof (Proof Sketch). 1. Fix $\varepsilon > 0$ such that the ε-ball B_ε around β does not meet the boundary $\partial \mathbb{R}_{\geq 0}^S$. By Lemma 1, outside B_ε, there exists $\delta_\varepsilon > 0$ such that the function $\mathcal{O}_k g < -\delta_\varepsilon$. Since $\mathcal{O}_k g$ is a continuous function of the specific rates k, a sufficiently small perturbation $\delta > 0$ in the rates will not change the sign of $\mathcal{O}_k g$. Hence, outside B_ϵ, the function g is strictly decreasing along trajectories $x(t)$ to Eq. 2. It follows that eventually every trajectory must enter B_ϵ.

2. Fix a sequence $\varepsilon_1 > \varepsilon_2 > \cdots > 0$ with ε_1 small enough so that the ε_1-ball around β does not meet the boundary $\partial \mathbb{R}_{\geq 0}^S$ and $\lim_{i \to \infty} \varepsilon_i \to 0$. For each ε_i, there exists j such that δ_j is small enough as per part (1) of the theorem. So every trajectory will eventually enter the ϵ_i neighborhood of β, and never leave. Since this is true for every i and $\lim_{i \to \infty} \varepsilon_i \to 0$, the result follows.

4   Main Result

The next definition makes precise our scheme, which takes a design matrix A to a reaction system $\mathcal{S}_{MLE}$ depending on A. The choice of this reaction system is not unique, but depends on two choices of basis. We proceed in two stages. In the first stage, we construct the reaction system $\mathcal{S}_{MLD}$ which solves the problem of finding the maximum likelihood distribution. In the second stage, we add reactions to solve for θ from the algebraic relations between the θ and X variables, obtaining $\mathcal{S}_{MLE}$.

Definition 8. *Fix a design matrix $A = (a_{ij})_{m \times n}$, a basis B for the free group $\mathbb{Z}^n \cap \ker A$, and a maximal linearly-independent subset B' of the columns of A.*

1. *The reaction network $\mathcal{R}_{MLD}(A, B)$ consists of n species $X_1, X_2, \ldots, X_n$ and for each $b \in B$, the reversible reaction:*

$$\sum_{j:b_j>0} b_j X_j \rightleftharpoons \sum_{j:b_j<0} -b_j X_j$$

2. *The reaction system $\mathcal{S}_{MLD}(A, B)$ consists of the reaction network $\mathcal{R}_{MLD}(A, B)$ with an assignment of rate 1 to each reaction.*
3. *The reaction network $\mathcal{R}_{MLE}(A, B, B')$ consists of $m+n$ species $\theta_1, \theta_2, \ldots, \theta_m$, $X_1, X_2, \ldots, X_n$, and in addition to the reactions in $\mathcal{R}_{MLD}$, the following reactions:*
   - *For each column $j \in B'$ of A, a reaction $\sum_{i=1}^m a_{ij}\theta_i \to 0$.*
   - *For each column $j \in B'$ of A, a reaction $X_j \to X_j + \sum_{i=1}^m a_{ij}\theta_i$.*
4. *The reaction system $\mathcal{S}_{MLE}(A, B, B')$ consists of the reaction network $\mathcal{R}_{MLE}(A, B, B')$ with an assignment of rate 1 to each reaction.*

Note that by the rank-nullity theorem of linear algebra, the dimension of the kernel plus the rank of the matrix equals the number of columns of the matrix. Hence counting the reversible reactions as two irreversible reactions, our scheme yields a reaction system whose number of reactions is twice the number of columns of A.

It is clear from the definition of $\mathcal{S}_{MLE}$ that the reactions that come from $\mathcal{R}_{MLD}$ are reversible and evolve without being affected by the other reactions. Hence we first prove global convergence of the reaction system $\mathcal{S}_{MLD}$ to the maximum likelihood distribution. This part is fairly straightforward. The key point is to verify that the reaction network $\mathcal{R}_{MLD}$ has no critical siphons. In fact, we show in the next lemma that $\mathcal{R}_{MLD}$ is prime, which will imply "no critical siphons" by Theorem 2.

Lemma 2. *Fix a design matrix $A = (a_{ij})_{m \times n}$ and a basis B for the free group $\mathbb{Z}^n \cap \ker A$. Then the reaction network $\mathcal{R}_{MLD}(A, B)$ is prime and $\mathcal{S}_{MLD}(A, B)$ is detailed balanced. Consequently, the reaction system $\mathcal{S}_{MLD}(A, B)$ is globally asymptotically stable.*

Proof. $\mathcal{R}_{MLD}(A, B)$ is prime by [17, Corollary 2.15]. The idea is to look at the toric model p_A as a ring homomorphism $\mathbb{C}[x_1, x_2, \ldots, x_n] \to \mathbb{C}[\mathbb{N}A]$ with $x_j \mapsto \theta^{a_j}$. (Here $\mathbb{N}A$ is the affine semigroup generated by the columns of A.) The kernel of this ring homomorphism is the associated ideal of $\mathcal{R}_{MLD}(A, B)$ by [17, Proposition 2.14], and the codomain is an integral domain, so the kernel must be prime.

Now $\mathcal{S}_{MLD}(A, B)$ is detailed balanced because the point $(1, 1, \ldots, 1) \in \mathbb{R}^n$ is a point of detailed balance since all rates are 1. Global asymptotic stability now follows from Theorems 2 and 3.

We can now obtain global convergence for $\mathcal{S}_{MLD}$.

Theorem 5 (The Reaction System $\mathcal{S}_{MLD}(A, B)$ Computes the Maximum Likelihood Distribution). *Fix a design matrix $A = (a_{ij})_{m \times n}$, a basis B for the free group $\mathbb{Z}^n \cap \ker A$, and a nonzero data vector $u \in \mathbb{Z}_{\geq 0}^n$. Let $x(t) = (x_1(t), x_2(t), \ldots, x_n(t))$ be a solution to the mass-action differential equations for the reaction system $\mathcal{S}_{MLD}(A, B)$ with initial conditions $x(0) = u/|u|_1$. Then $x(\infty) := \lim_{t \to \infty} x(t)$ exists and equals the maximum likelihood distribution $\hat{p}(u)$.*

Proof. For the system $\mathcal{S}_{MLD}(A, B)$, note that $(x(0) + H) \cap \mathbb{R}_{>0}^n = P_A(u/|u|_1)$. By Theorem 3, $x(\infty)$ exists, and the function $\sum_{i=1}^n x_i \log x_i - x_i - x_i \log 1$ attains its unique minimum in $P_A(u/|u|_1)$ at $x(\infty)$. Since the system is mass-conserving, $\sum_{i=1}^n x_i$ is constant on $P_A(u/|u|_1)$, so this is equivalent to the fact that Shannon entropy $H(x) = -\sum_{i=1}^n x_i \log x_i$ is increasing, and attains its unique maximum value in $P_A(u/|u|_1)$ at $x(\infty)$. By Theorem 1, the point $x(\infty)$ must be the maximum likelihood distribution $\hat{p}(u)$.

As the reversible reactions in $\mathcal{S}_{MLE}$ approach closer and closer to equilibrium, we wish to absorb the values of the X variables into reaction rates and pretend that the irreversible reactions are reactions only in the θ variables. This has the advantage that we can treat this pretend reaction system in the θ variables as a perturbation of a reversible, detailed balanced system. We can then hope to employ Theorem 4 and conclude global convergence for these irreversible reactions, and hence for $\mathcal{S}_{MLE}$.

One small technical point deserves mention. The pretend reaction system in the θ variables is not a reaction system since the rates are not real numbers but functions of time. This will not trouble us. We have already provisioned for this in Definition 7 by allowing perturbations of reaction systems to be differential inclusions.

Theorem 6 (The Reaction System $\mathcal{S}_{MLE}(A, B, B')$ Computes the Maximum Likelihood Estimator). *Fix a design matrix $A = (a_{ij})_{m \times n}$, a basis B for the free group $\mathbb{Z}^n \cap \ker A$, and a nonzero data vector $u \in \mathbb{Z}_{\geq 0}^n$. Let $x(t) = (x_1(t), x_2(t), \ldots, x_n(t), \theta_1(t), \theta_2(t), \ldots, \theta_m(t))$ be a solution to the mass-action differential equations for the reaction system $\mathcal{S}_{MLE}(A, B, B')$ with initial conditions $x(0) = u/|u|_1$ and $\theta(0) = 0$. Then $x(\infty) := \lim_{t \to \infty} x(t)$ exists and equals the maximum likelihood distribution $\hat{p}(u)$, and $\theta(\infty) := \lim_{t \to \infty} \theta(t)$ exists and equals the maximum likelihood estimator $\hat{\theta}(u)$.*

Proof (Proof Sketch). Fix u and let $\hat{p} = \hat{p}(u)$ and $\hat{\theta} = \hat{\theta}(u)$. Note that for the species $X_1, X_2, \ldots, X_n$, the differential equations for the systems $\mathcal{S}_{MLE}(A, B)$ and $\mathcal{S}_{MLD}(A, B, B')$ are identical, since these species appear purely catalytically in the reactions that belong to $\mathcal{R}_{MLE}(A, B, B') \backslash \mathcal{R}_{MLD}(A, B)$. Hence $x(\infty) = \hat{p}(u)$ follows from Theorem 5.

To see that $\theta(\infty) = \hat{\theta}$, let us first allow the X species to reach equilibrium, then treat the θ system with replacing the X species by rate constants representing their values at equilibrium. The system $\Theta_{MLE}(A, B, B', x(\infty))$ obtained in this way in only the θ species is a reaction system with the reactions

– For each column $j \in B'$ of A, a reaction $\sum_{i=1}^{m} a_{ij}\theta_i \to 0$ of rate 1
– For each column $j \in B'$ of A, a reaction $0 \to \sum_{i=1}^{m} a_{ij}\theta_i$ of rate $x_j(\infty)$.

This is a reversible reaction system, and the maximum likelihood estimators $\hat{\theta}$ are precisely the points of detailed balance for this system, where we are using the fact that B' was a maximal linearly-independent set of the columns of A. In addition, this system has no siphons since if species θ_i is absent, and $a_{ij} > 0$ then θ_i will immediately be produced by the reaction $0 \to \sum_{i'=1}^{m} a_{i'j}\theta_{i'}$. (We are assuming A has no 0 row. If A has a 0 row, we can ignore it anyway.) It follows from Theorem 3 that this system is globally asymptotically stable, and every trajectory approaches a maximum likelihood estimator $\hat{\theta}$.

We claim that trajectories of our actual system are also trajectories of a perturbation of the system $\Theta_{MLE}(A, B, B', x(\infty))$. Consider any trajectory $(x(t), \theta(t))$ to $\mathcal{S}_{MLE}(A, B, B')$ starting at $(u/|u|_1, 0)$. We are going to consider the projected trajectory $(\mathbb{R}_\geq, \theta)$. We now show that it is possible to choose appropriate t_i and δ_i so that $((t_i, \infty), \theta(t))$ is a trajectory of a δ_i-perturbation of $\Theta_{MLE}(A, B, B', x(\infty))$, for $i = 1, 2, \ldots$.

Wait for a sufficiently large time t_1 till $x(t)$ is in a sufficiently small δ_1 neighborhood of $x(\infty)$ which it will never leave. After this time, we obtain a differential inclusion in the θ species with the mass-action equations 2 for the reactions

– For each column j of A, a reaction $\sum_{i=1}^{m} a_{ij}\theta_i \to 0$ of rate 1
– For each column j of A, a reaction $0 \to \sum_{i=1}^{m} a_{ij}\theta_i$ with time-varying rate lying in the interval $(x_j(\infty) - \delta_1, x_j(\infty) + \delta_1)$.

Continuing in this way, we choose a decreasing sequence $\delta_1 > \delta_2 > \cdots > 0$ with $\lim_{i \to \infty} \delta_i \to 0$, and corresponding times $t_1 < t_2 < t_3 \ldots$ with $\lim_{i \to \infty} t_i \to \infty$ such that after time t_i, $x(t)$ is in a δ_i neighborhood of $x(\infty)$ which it will never leave. Then $((t_i, \infty), \theta(t))$ is a trajectory of the δ_i-perturbation of the system $\Theta_{MLE}(A, B, B', x(\infty))$. Hence $\theta(t)$ satisfies the conditions of Theorem 4. Hence $\lim_{t \to \infty} \theta(t) = \hat{\theta}$.

5   Related Work and Conclusions

The mathematical similarities of both log-linear statistics and reaction networks to toric geometry have been pointed out before [9,17]. Craciun et al. [9] refer to the steady states of complex-balanced reaction networks as *Birch points* "to highlight the parallels" with algebraic statistics. This paper develops on these observations, and serves to flesh out this mathematical parallel into a scheme for molecular computation.

Various building blocks for molecular computation that assume mass-action kinetics have been proposed before. We briefly review some of these proposals.

In [19], Napp and Adams model molecular computation with mass-action kinetics, as we do here. They propose a molecular scheme to implement message passing schemes in probabilistic graphical models. The goal of their scheme is to convert a factor graph into a reaction network that encodes the single-variable marginals of the joint distribution as steady state concentrations. In comparison, the goal of our scheme is to do statistical inference and compute maximum likelihood estimators for log-linear models. Napp and Adams focus on the "forward model" task of how a given data-generating process (a factor graph) can lead to observed data, whereas our focus is on the "backward model" task of inference, going from the observed data to the data-generating process. Further our scheme couples the deep role that MaxEnt algorithms play in Machine Learning with MaxEnt's roots in the Second Law of Thermodynamics whereas Napp and Adams are drawing their inspiration from variable elimination implemented via message passing which has its roots in Boolean constraint satisfaction problems.

Qian and Winfree [23,24] have proposed a DNA gate motif that can be composed to build large circuits, and have experimentally demonstrated molecular computation of a Boolean circuit with around 30 gates. In comparison, our scheme natively employs a continuous-time dynamical system to do the computation, without a Boolean abstraction.

Taking a control theory point of view, Oishi and Klavins [20] have proposed a scheme for implementing linear input/output systems with reaction networks. Note that for a given matrix A, the set of maximum likelihood distributions is usually not linear, but log-linear.

Daniel et al. [10] have demonstrated an in vivo implementation of feedback loops, exploiting analogies with electronic circuits. It is possible that the success of their schemes is also related to the toric nature of mass-action kinetics.

Buisman et al. [5] have proposed a reaction network scheme for computation of algebraic functions. The part of our scheme which reads out the maximum likelihood estimator from the maximum likelihood distribution bears some similarity to their work.

One limitation of our present work is that the number of columns of the matrix A can become very large, for example $2^{|V|}$ for a graphical model with V nodes. Since the number of species and number of reactions both depend on the number of columns of A, this can require an exponentially large reaction network which may become impractical. One direction for future work is to extend our scheme by specifying a reaction network that computes maximum likelihood for graphical models.

We have some freedom in our scheme in the choice of basis sets B and B'. In any chemical implementation of this work, there might be opportunity for optimization in choice of basis.

Acknowledgements. I thank Nick S. Jones, Anne Shiu, Abhishek Behera, Ezra Miller, Thomas Ouldridge, Gheorghe Craciun, and Bence Melykuti for useful discussions.

References

1. Agresti, A.: Categorical Data Analysis. Wiley, New York (2013)
2. Angeli, D., De Leenheer, P., Sontag, E.D.: A Petri net approach to the study of persistence in chemical reaction networks. Math. Biosci. **210**(2), 598–618 (2007)
3. Benenson, Y., Gil, B., Ben-Dor, U., Adar, R., Shapiro, E.: An autonomous molecular computer for logical control of gene expression. Nature **429**(6990), 423–429 (2004)
4. Bishop, Y.M.M., Feinberg, S., Holland, P.: Discrete Multivariant Analysis. The MIT Press, Cambridge (1975)
5. Buisman, H.J., ten Eikelder, H.M.M., Hilbers, P.A.J., Liekens, A.M.L.: Computing algebraic functions with biochemical reaction networks. Artif. Life **15**(1), 5–19 (2009)
6. Cardelli, L.: Strand algebras for DNA computing. Nat. Comput. **10**, 407–428 (2011)
7. Cardelli, L.: Two-domain DNA strand displacement. Math. Struct. Comput. Sci. **23**(02), 247–271 (2013)
8. Christensen, R.: Log-Linear Models and Logistic Regression, vol. 168. Springer, New York (1997)
9. Craciun, G., Dickenstein, A., Shiu, A., Sturmfels, B.: Toric dynamical systems. J. Symb. Comput. **44**(11), 1551–1565 (2009). In Memoriam Karin Gatermann
10. Daniel, R., Rubens, J.R., Sarpeshkar, R., Lu, T.K.: Synthetic analog computation in living cells. Nature **497**(7451), 619–623 (2013)
11. Doyle, J., Csete, M.: Rules of engagement. Nature **446**(7138), 860–860 (2007)
12. Fienberg, S.E., Rinaldo, A., et al.: Maximum likelihood estimation in log-linear models. Ann. Stat. **40**(2), 996–1023 (2012)
13. Gopalkrishnan, M.: Catalysis in reaction networks. Bull. Math. Biol. **73**, 2962–2982 (2011). doi:10.1007/s11538-011-9655-3
14. Gopalkrishnan, M., Miller, E., Shiu, A.: A geometric approach to the global attractor conjecture. In preparation
15. Horn, F.J.M.: The dynamics of open reaction systems. In: Mathematical Aspects of Chemical and Biochemical Problems and Quantum Chemistry. Proceedings of the SIAM-AMS Symposium Applied Mathematics, vol. 8, New York (1974)
16. Lauritzen, S.L.: Graphical Models. Oxford University Press, Oxford (1996)
17. Miller, E.: Theory and applications of lattice point methods for binomial ideals. In: Fløystad, G., Johnsen, T., Knutsen, A.L. (eds.) Combinatorial Aspects of Commutative Algebra and Algebraic Geometry. Abel Symposia, vol. 6, pp. 99–154. Springer, Heidelberg (2011)
18. Milo, R., Shen-Orr, S., Itzkovitz, S., Kashtan, N., Chklovskii, D., Alon, U.: Network motifs: simple building blocks of complex networks. Science **298**(5594), 824–827 (2002)
19. Napp, N.E., Adams, R.P.: Message passing inference with chemical reaction networks. In: Advances in Neural Information Processing Systems, pp. 2247–2255 (2013)
20. Oishi, K., Klavins, E.: Biomolecular implementation of linear I/O systems. IET Syst. Biol. **5**(4), 252–260 (2011)

21. Pachter, L., Sturmfels, B.: Algebraic Statistics for Computational Biology. Algebraic Statistics for Computational Biology, vol. 13. Cambridge University Press, New York (2005)
22. Qian, L., Soloveichik, D., Winfree, E.: Efficient turing-universal computation with DNA polymers. In: Sakakibara, Y., Mi, Y. (eds.) DNA 16 2010. LNCS, vol. 6518, pp. 123–140. Springer, Heidelberg (2011)
23. Qian, L., Winfree, E.: A simple DNA gate motif for synthesizing large-scale circuits. J. R. Soc. Interface (2011)
24. Qian, L., Winfree, E.: Scaling up digital circuit computation with DNA strand displacement cascades. Science **332**(6034), 1196–1201 (2011)
25. Shapiro, E., Benenson, Y.: Bringing DNA computers to life. Sci. Am. **294**(5), 44–51 (2006)
26. Shinar, G., Feinberg, M.: Structural sources of robustness in biochemical reaction networks. Science **327**(5971), 1389–1391 (2010)
27. Soloveichik, D., Seelig, G., Winfree, E.: DNA as a universal substrate for chemical kinetics. Proc. Natl. Acad. Sci. **107**(12), 5393–5398 (2010)
28. Sontag, E.D.: Structure and stability of certain chemical networks and applications to the kinetic proofreading model of T-cell receptor signal transduction. IEEE Trans. Autom. Control **46**, 1028–1047 (2001)
29. Thomson, M., Gunawardena, J.: Unlimited multistability in multisite phosphorylation systems. Nature **460**(7252), 274–277 (2009)
30. Tyson, J.J., Chen, K.C., Novak, B.: Sniffers, buzzers, toggles and blinkers: dynamics of regulatory and signaling pathways in the cell. Current Opin. Cell Biol. **15**(2), 221–231 (2003)
31. Yordanov, B., Kim, J., Petersen, R.L., Shudy, A., Kulkarni, V.V., Phillips, A.: Computational design of nucleic acid feedback control circuits. ACS Synth. Biol. **3**(8), 600–616 (2014)

Nondeterministic Seedless Oritatami Systems and Hardness of Testing Their Equivalence

Yo-Sub Han[1], Hwee Kim[1(✉)], Makoto Ota[2], and Shinnosuke Seki[2]

[1] Department of Computer Science, Yonsei University, 50 Yonsei-Ro,
Seodaemum-Gu, Seoul 120-749, Republic of Korea
{emmous,kimhwee}@yonsei.ac.kr
[2] University of Electro-Communications, 1-5-1,
Chofugaoka, Chofu, Tokyo 182-8585, Japan
o1111032@edu.cc.uec.ac.jp, s.seki@uec.ac.jp

Abstract. The oritatami system (OS) is a model of computation by cotranscriptional folding, being inspired by the recent experimental succeess of RNA origami to self-assemble an RNA tile cotranscriptionally. The OSs implemented so far, including binary counter and Turing machine simulator, are deterministic, that is, uniquely fold into one conformation, while nondeterminism is intrinsic in biomolecular folding. We introduce nondeterminism to OS (NOS) and propose an NOS that chooses an assignment of Boolean values nondeterministically and evaluates a logical formula on the assignment. This NOS is seedless in the sense that it does not require any initial conformation to begin with like the RNA origami. The NOS allows to prove the co-NP hardness of deciding, given two NOSs, if there exists no conformation that one of them folds into but the other does not.

1 Introduction

In nature, an one-dimensional RNA sequence—a primary structure—folds itself autonomously and forms a more complex secondary structure. It has been a constant question to predict the secondary structure from a given primary structure, and based on experimental observations, researchers established various RNA secondary structure prediction models including RNAfold [12], Pknots [9], mFold [11] and KineFold [10]. Traditional models tend to rely on the energy optimization of the whole structure.

Recently, biochemists showed that the kinetics—the step-by-step dynamics of the reaction—plays an essential role in the geometric shape of the RNA foldings [2], since the folding caused by intermolecular reactions is faster than the RNA transcription rate [7]. By controlling cotranscriptional foldings, researchers succeeded in cotranscriptionally assembling a rectangular tile out of RNA, which is called RNA Origami [5] as depicted in Fig. 1. From this kinetic point of view, Geary et al. [4] proposed a new folding model called the oritatami system (OS). In general, OS defines a sequence of beads (which is the primary structure) and a set of rules for possible intermolecular reactions between beads. For each bead

© Springer International Publishing Switzerland 2016
Y. Rondelez and D. Woods (Eds.): DNA 2016, LNCS 9818, pp. 19–34, 2016.
DOI: 10.1007/978-3-319-43994-5_2

Fig. 1. RNA Origami [5]. The artwork is by Cody Geary.

in the sequence, the system takes a lookahead of a few upcoming beads and determines the best location of the bead that maximizes the number of possible interactions from the lookahead. Note that the lookahead represents the reaction rate of the cotranscriptional folding and the number of interactions represents the energy level. In OS, we call the secondary structure *the conformation*, and the resulting secondary structure *the terminal conformation*.

Geary et al. implemented an OS to count in binary [3] and an OS to simulate a cyclic tag system [4]. These OSs uniquely folds into one conformation, and in this sense, they are deterministic. In contrast, nondeterminism is intrinsic in biomolecular folding. Therefore, we define the nondeterministic OS (NOS) in this paper, and examine its power. It turns out that nondeterminism can be made use of for OSs to execute randomized algorithms. We propose an NOS that evaluates a Boolean formula in disjunctive normal form (DNF formula) on a random assignment. This NOS is in fact seedless like the RNA origami. More importantly, the NOS proves the co-NP hardness of the OSEQ problem, which asks, given two NOSs, if there exists no conformation that one folds into but the other does not (Theorem 2). The equivalence test is indispensable in the optimization of the design of a given OS. As we will see, the equivalence of two deterministic OSs is testable in linear time (Theorem 1).

2   Preliminaries

Let Σ be a set of bead types, and Σ^* be the set of finite strings of beads, i.e., strings over Σ, including the empty string λ. Let $w = a_1 a_2 \cdots a_n$ be a string of length n for some integer n and bead types $a_1, \ldots, a_n \in \Sigma$. The *length* of w is denoted by $|w|$. For two indices i, j with $1 \leq i \leq j \leq n$, we let $w[i, j]$ refer to the substring $a_i a_{i+1} \cdots a_{j-1} a_j$; if $i = j$, then we denote it by $w[i]$ instead. We use w^n to denote the string $\underbrace{ww \cdots w}_{n}$.

Oritatami systems operate on the hexagonal lattice. The *grid graph* of the lattice is the graph whose vertices correspond to the lattice points and connected if the corresponding lattice points are at unit distance hexagonally. For a point p and a bead type $a \in \Sigma$, we call the pair (p, a) an *annotated point*, or simply a *point* if being annotated is clear from context. Two annotated points $(p, a), (q, b)$ are *adjacent* if pq is an edge of the grid graph.

A *path* is a sequence $P = p_1 p_2 \cdots p_n$ of *pairwise-distinct* points $p_1, p_2, \ldots, p_n$ such that $p_i p_{i+1}$ is at unit distance for all $1 \leq i < n$. Given a string $w \in \Sigma^*$ of bead types of length n, a *path annotated by* w, or simply w-*path*, is a sequence P_w of annotated points $(p_1, w[1]), \ldots, (p_n, w[n])$, where $p_1 \cdots p_n$ is a path. Annotated points of the w-path are regarded as a bead, and hence, we call them beads and, in particular, we call the i-th point $(p_i, w[i])$ the i-th bead of the w-path.

Let $\mathcal{H} \subseteq \Sigma \times \Sigma$ be a symmetric relation, specifying between which types of beads can form a hydrogen-bond-based interaction (h-interaction for short). This relation $\mathcal{H}$ is called the *ruleset*. It is convenient to assume a special *inert* bead type $\bullet \in \Sigma$ that never forms any h-interaction according to $\mathcal{H}$.

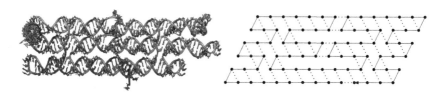

Fig. 2. (Left) An example of an RNA tile generated by RNA Origami. (Right) A conformation representing the RNA tile in OS. The directed solid line represents a path, dots represent beads, and dotted lines represent h-interactions. The idea and artwork were provided by Cody Geary.

A *conformation* C is a pair of a w-path $P_w = (p_1, w[1])(p_2, w[2]) \cdots$ and a set H of h-interactions, where $H \subseteq \{\{i,j\} \mid 1 \le i, i + 2 \le j\}$ and $\{i, j\} \in H$ implies that the i-th and j-th beads of the w-path form an h-interaction between them. An example conformation is found in Fig. 2 (Right). The condition $i+2 \le j$ represents the topological restriction that two beads $(p_i, w[i]), (p_{i+1}, w[i+1])$, adjacent to each other on the w-path, cannot form an h-interaction between them. We say C is *finite* if its path is finite. From now on, when a conformation is illustrated, any unlabeled bead is assumed to be labeled with $\bullet$, that is, be inert. For an integer $\alpha \ge 1$, C is of *arity* α if none of its beads interact with more than α beads. On the hexagonal lattice where every point is adjacent to six points, $\alpha > 6$ is merely meaningless, but on another lattice larger α's may. Let $\mathcal{C}_\alpha$ be the set of all conformations of arity-α.

A rule (a, b) in the ruleset $\mathcal{H}$ is *used* in the conformation C if there exists $\{i, j\} \in H$ such that $w[i] = a$ and $w[j] = b$ or $w[i] = b$ and $w[j] = a$. A conformation C is *valid (with respect to $\mathcal{H}$)* if for all $\{i, j\} \in H$, $(w[i], w[j]) \in \mathcal{H}$. In a context with one fixed ruleset, only valid conformations with respect to the ruleset are considered, and we may not specify with respect to what ruleset they are valid.

Given a ruleset $\mathcal{H}$ and a valid finite conformation $C_1 = (P_w, H)$ with respect to $\mathcal{H}$, we say that another conformation C_2 is an *elongation* of C_1 by a bead $a \in \Sigma$ if $C_2 = (P_w \cdot (p, a), H \cup H')$ for some lattice point p and (possibly empty) set of h-interactions $H' \subseteq \{\{i, |w|+1\} \mid 1 \le i \le |w|, (w[i], a) \in \mathcal{H}\}$. Note that C_2 is also valid. For a conformation C and a finite string $w \in \Sigma^*$, by $\mathcal{E}(C, w)$, we denote the set of all elongations of C by w, that is, $\mathcal{E}(C, w) = \{C' \in \mathcal{C} \mid C \to_w^* C'\}$. For an arity α, let $\mathcal{E}_\alpha(C, w) = \mathcal{E}(C, w) \cap \mathcal{C}_\alpha$.

2.1   Oritatami System

An *oritatami system* (OS) is a 5-tuple $\Xi = (\mathcal{H}, \alpha, d, \sigma, w)$, where $\mathcal{H}$ is a ruleset, α is an *arity*, $d \geq 1$ is a positive integer called the *delay*, σ is an initial valid conformation of arity α called the *seed*, and w is a possibly-infinite string of beads called a *primary structure*.

The delay d, arity-α oritatami system Ξ cotranscriptionally folds its primary structure in the following way. For a string $x \in \Sigma^*$, a conformation C_1, and an elongation C_2 of C_1 by $x[1]$, we say that Ξ *(cotranscriptionally) folds x upon C_1 into C_2* if

$$C_2 \in \underset{C \in \mathcal{E}_\alpha(C_1, x[1])}{\mathrm{argmin}} \; \min \left\{ \Delta G(C') \mid C' \in \mathcal{E}_\alpha(C, x[2, k])), 2 \leq k \leq d \right\}, \qquad (1)$$

where $\Delta G(C')$ is an energy function that assigns C' with the negation of the number of h-interactions within C' as energy. Informally speaking, C_2 is a conformation obtained by elongating C_1 by the bead $x[1]$ so as for the beads $x[1], x[2], \ldots, x[d]$ to create as many h-interactions as possible. Then we write $C_1 \overset{\Xi}{\hookrightarrow}_x C_2$, and the superscript Ξ is omitted whenever Ξ is clear from context. Through the folding, the first bead of x is *stabilized*. In figures, we conventionally color x—the fragment to be stabilized—in cyan.

Example 1 (Glider). Let us explain how the OS cotranscriptionally folds a motif called the *glider*. Gliders offer a directional linear conformation and have been heavily exploited in the existing studies on OS [3]. Consider a delay-3 OS whose seed is the black conformation in Fig. 3 (a), primary structure is $w = b \bullet ac \bullet bd \bullet c \cdots$, and the ruleset is $\mathcal{H} = \{(a, a), (b, b), (c, c), (d, d)\}$.

By the fragment $w[1, 3] = b \bullet a$, the seed can be elongated in many ways; three of them are shown in Fig. 3 (a). The only bead on the fragment that may form a new h-interaction is a (b is also capable according to $\mathcal{H}$ but no other b is around). In order for the a to get next to the other a, on the seed, the b on the fragment must be located to the east of the last bead of the seed; thus, the b is stabilized there, as shown in Fig. 3 (b). The stabilization transcribes the next bead $w[4] = c$. The sole other c around is on the seed but is too far for the c just transcribed to get adjacent to. Thus, the only way for the fragment $w[2, 4] = \bullet ac$ to form an h-interaction is to put the two a's next to each other as before, and for that, the $\bullet$ must be located to the southeast of the preceding b as shown in Fig. 3 (c). The next bead to be transcribed, $w[5]$, is inert, and hence, cannot override the previous decision to put the two a's next to each other. The first six beads have been thus stabilized as shown in Fig. 3 (d), and the glider has thus moved forward by distance 2.

It is easily induced inductively that gliders of arbitrary "flight distance" d can be folded by a delay-3 OS; such long-distance gliders have been used in [3]. Moreover, as suggested in Fig. 3, a constant number of bead types are enough for that (in this example, $a, b, c, d, \bullet$).

Gliders also provide a medium to propagate 1-bit information at arbitrary distance. The height (up or down) of the first bead determines whether the

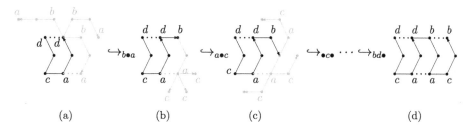

Fig. 3. A glider folded by a delay-3 OS. (a) Three ways to elongate the current conformation by the fragment $b \bullet a$ among many. (b) Three most stable elongations by the fragment $\bullet ac$. (c) Three ways to elongate the current conformation by the fragment $ac\bullet$ among many. (d) The stabilization of $b \bullet ac \bullet b$.

last bead is stabilized up or down after the glider traverses the distance d. For instance, the glider in Fig. 3 launches up and thus its last bead (the second b) also comes up after traveling the distance $d = 2$; being launched down implies being terminated down. This capability has been exploited for an OS to simulate a cyclic tag system for Turing universality [3] and the NOS that we shall propose in Sect. 3 also uses it.   □

Note that cotranscriptional folding, as formulated in (1), considers not only elongations of C by $x[2, d]$ but also those by prefixes of $x[2, d]$, that is, $x[2], x[2, 3]$, ..., $x[2, d-1]$. This is necessary to fully fold the primary structure till the end or when there is not enough space around the last bead of C to elongate C by the whole $x[2, d]$ (See Fig. 4.). Otherwise, under the current energy function, the optimization just ignores those "halfway" elongations because more beads never rise energy. Under other "more realistic" energy functions, halfway elongations would play a more active role in the folding.

The set $\mathcal{F}(\varXi)$ of all conformations foldable by $\varXi$ is now defined recursively as follows: $\sigma \in \mathcal{F}(\varXi)$, and if $C_i \in \mathcal{E}_\alpha(\sigma, w[1, i])$ is in $\mathcal{F}(\varXi)$ and $C_i \stackrel{\varXi}{\hookrightarrow}_{w[i+1, i+d]} C_{i+1}$, then $C_{i+1} \in \mathcal{F}(\varXi)$. A finite conformation $C_i \in \mathcal{E}_\alpha(\sigma, w[1, i])$ foldable by $\varXi$ is *terminal* if one of the following conditions holds:

1. the primary structure w is finite and $i = |w|$;
2. either w is infinite or $i < |w|$, and there exists no conformation C_{i+1} such that $C_i \stackrel{\varXi}{\hookrightarrow}_{w[i+1, i+\delta]} C_{i+1}$.

Note that not only the conformation in Fig. 4 (d) but also the conformation in Fig. 4 (f) is terminal by the second condition of the terminal conformation. By $\mathcal{F}_\square(\varXi)$, we denote the set of all terminal conformations foldable by $\varXi$.

The OS $\varXi$ is *deterministic* if any foldable conformation $C_i \in \mathcal{E}_\alpha(\sigma, w[1, i])$ is either terminal or admits a unique conformation C_{i+1} such that $C_i \stackrel{\varXi}{\hookrightarrow}_{w[i+1, i+\delta]}$ C_{i+1}, that is, every bead is stabilized uniquely. For example, the system in Fig. 4 is nondeterministic. Note that nondeterministic systems fold into multiple terminal conformations as suggested in Fig. 4. On the other hand, deterministic

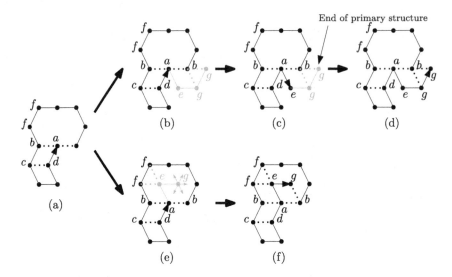

Fig. 4. The two cases in which cotranscriptional folding considers "halfway" elongations. (b) and (e) show two most stable elongations of the current conformation (a); the one in (e) is a halfway elongation. (b) to (d) show the case when folding is almost over, and (e) and (f) show the case when there is not enough space. The conformations in (d) and (f) are both terminal, though the one in (f) is "shorter."

systems fold into exactly one terminal conformation by definition. Thus, an oritatami system is deterministic if and only if the system folds into one terminal conformation. The property of being deterministic is decidable in linear time. Indeed, it suffices to run an OS and checks whether it encounters nondeterminism or not.

Example 2 (Assignor). Let us exhibit here how nondeterminism is used in the OS that we shall propose in Sect. 3, or more particularly, in its module called the *assignor*. The OS is of delay 3, with a ruleset including $(10^{eb}, 3^a), (10^{eb}, 9^a), (12^{eb}, 3^{eb}), (12^{eb}, 9^a), (12^{eb}, 4^a)$. The OS folds the assignor uniquely as shown in Fig. 5 (a), up to its fourth last bead. The last three beads of the assignor are 10^{eb}, 11^{eb} and 12^{eb}.

The fragment 10^{eb}-11^{eb}-12^{eb} can be fold in two ways equally stably with three h-interactions as shown in Fig. 5 (b-1) and (b-2), and no more no matter how the fragment is folded. Hence, the bead 10^{eb} is stabilized in two ways as shown in Fig. 5 (c-1) and (c-2). The remaining beads 11^{eb} and 12^{eb} are stabilized uniquely one after another as shown in Fig. 5 (d-1) and (d-2). The assignor nondeterministically stabilizes the last bead 12^{eb} up or down. In our NOS, this random assignment of 1-bit information is propagated by gliders in the way mentioned in Example 1. □

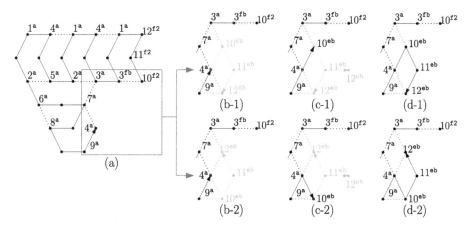

Fig. 5. An assignor folded by a delay 3 OS. While stabilizing the bead 10^{eb}, two elongations equally give three interactions and the bead nondeterministically stabilizes at two different points. (a) The conformation up to the fourth last bead. (b) Two ways to elongate the current conformation by the fragment 10^{eb}-11^{eb}-12^{eb}. (c) Ways to elongate the current conformation by the fragment 11^{eb}-12^{eb}. (d) Two final conformations.

We say that an OS is *seedless* if its seed is $(\lambda, \emptyset)$. A seedless OS can start folding at any point of the lattice. If a conformation C is foldable, then any of its congruence, that is, a conformation obtained by applying a combination of translation, rotation, and reflection to C is also foldable. Therefore, fixing the first bead to the origin of the lattice and the second bead to one specific neighbor of the origin does not cause any loss of generality. In this sense, we can regard an OS with a seed of at most 2 beads seedless. Furthermore, a seed of 3 beads can make an OS seedless. Imagine an OS whose seed consists of 2 beads. if an elongation of the seed by the first bead is foldable, then its reflection across the seed, which is just a line segment, is also foldable. Hence, if the first bead is stabilized uniquely, then it can be rather considered as a part of the seed. That is the case for the OS we shall propose in Sect. 3. In this sense, the OS we propose is seedless.

We formally define the equivalence of two oritatami systems.

Definition 1. *Two oritatami systems Ξ_1 and Ξ_2 are equivalent if $\mathcal{F}_\square(\Xi_1) = \mathcal{F}_\square(\Xi_2)$, namely, if there is no terminal conformation that one folds into but the other does not.*

Then, we define oritatami system equivalence test.

Definition 2. *Given two oritatami systems Ξ_1 and Ξ_2, oritatami system equivalence test (OSEQ) is to determine whether or not they are equivalent.*

Since it takes linear time to simulate a deterministic OS, we establish the following theorem.

Theorem 1. *For two deterministic oritatami systems, the OSEQ problem can be solved in linear time.*

On the other hand, the problem for NOSs turns out to be hard, and we shall prove the following theorem in Sect. 3.

Theorem 2. *The OSEQ problem is co-NP complete, even if the two input NOSs differ only in ruleset, and their rulesets $\mathcal{H}_1, \mathcal{H}_2$ satisfy $\mathcal{H}_1 = \mathcal{H}_2 \cup \{(a, b)\}$ for some rule (a, b).*

3   Seedless NOS as a DNF Verifier

We propose a seedless NOS that evaluates a given DNF formula, i.e., a Boolean formula in the disjunctive normal form, and then make use of it to prove the coNP-hardness of deciding if two given NOSs are equivalent even under a severe constraint.

A DNF formula ϕ is written as $\bigvee_{1 \le i \le n} C_i$ for some clauses $C_1, \ldots, C_n$ that is a logical AND ($\wedge$) of some of the Boolean variables $v_1, \ldots, v_m$ and their negations. The *DNF tautology problem* asks if a given DNF formula is evaluated to TRUE on all possible assignments. The problem is coNP-hard, since it can be polynomially reduced from the dual problem of the satisfiability problem, which is NP-complete [6].

Algorithm 1. Evaluate a DNF formula with m variables and n clauses formula on a randomly chosen assignment

---

1 **for** $k = 1$ **to** n **do** $c[k] \leftarrow *$;
2 **for** $i = 1$ **to** m **do**
3 $\quad$ Randomly assign TRUE or FALSE into v_i.;
4 $\quad$ **for** $k = 1$ **to** n **do**
5 $\quad\quad$ **if** *The k-th clause involves v_i* **and** $v_i =$ FALSE **then** $c[k] \leftarrow$ U;
6 $\quad\quad$ **else if** *The k-th clause involves $\neg v_i$* **and** $v_i =$ TRUE **then** $c[k] \leftarrow$ U;

7 **for** $k = 1$ **to** n **do**
8 $\quad$ **if** $c[k] = *$ **then return** Satisfied;

9 **return** *Unsatisfied*

Algorithm 1 evaluates the DNF formula ϕ on a random assignment. For the ease of explanation, we assume m is even (otherwise we just assume one more imaginary variable that occurs nowhere). The seedless NOS Ξ_τ evaluates ϕ using this algorithm. Both its delay and arity are set to 3 (in fact, the arity can be set to any value larger). We conventionally use the term *context* to denote beads and interactions around the current bead that we consider during the stabilization.

Its primary structure is of the form $w_s w_1 w_2 \cdots w_m w_v$. We call the factors $w_s, w_1, \ldots, w_m, w_v$ *modules*. The NOS is to fold the primary structure in

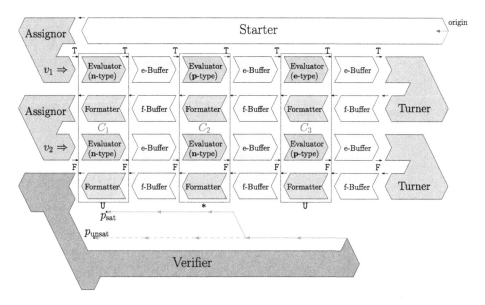

Fig. 6. An overview of the NOS that evaluates a given DNF formula $\phi = (\neg v_1 \wedge \neg v_2) \vee (v_1 \wedge \neg v_2) \vee (v_2)$, with two variables v_1, v_2 and three clauses C_1, C_2, C_3, on the assignment $(v_1, v_2) = (\mathtt{T}, \mathtt{F})$ that it chooses nondeterministically. The folding starts from the purple arrow in the starter. In the last module called the verifier, the conformation folds to p_{sat} since the formula is satisfied. Pink dashed lines represent an alternate conformation that stops at p_{unsat} when the formula is not satisfied. (Color figure online)

a zigzag manner as outlined in Fig. 6. The first module w_s named the *starter* folds into the glider as shown in Fig. 7 and offers a linear scaffold of width $O(n)$, which serves as a "seed" for the succeeding modules. This module admits no other conformation, and it is almost straightforward to design a module that folds uniquely by hardcording the target conformation into the primary structure and ruleset. In fact, all the rules used in the glider in Fig. 7 are sufficient (and necessary) for the primary structure of w_s to fold into the glider. Note that we use superscripts to indicate sets of bead types used for different modules, i.e., $\mathtt{f2}$ is used for formatters and $\mathtt{s}$ is used for starters. Being thus folded, the starter exposes below n copies of the sequence of bead types $10^{\mathtt{f2}}\text{-}9^{\mathtt{f2}}\text{-}8^{\mathtt{f2}}\text{-}7^{\mathtt{f2}}$ at a fixed interval (every 8 points), which shall be translated by succeeding modules as the corresponding clause being satisfiable (denoted by $*$). This corresponds to the initialization of the array c in line 1 of Algorithm 1.

The next module w_1 consists of submodules. The first submodule is the assignor explained in Example 2. It nondeterministically stabilizes its last bead up or down, and the OS interprets up as $\mathtt{TRUE}$ being assigned to v_1 and down as false being assigned to v_1 (See Fig. 6, where $\mathtt{TRUE}$ is assigned to v_1, for instance.). The assignor is succeeded by submodules called evaluators and buffers, which occur alternately. The buffer is just a glider. There are n evaluators in w_1; one for each of the n clauses. The k-th evaluator, for C_k, takes the value (T/F) assigned

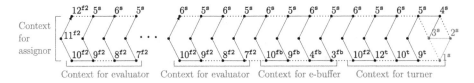

Fig. 7. The linear scaffold conformation into which the starter w_s deterministically folds. Three blue beads indicate the seed. (Color figure online)

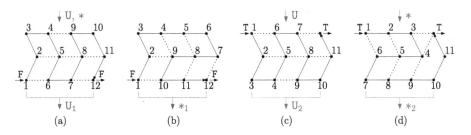

Fig. 8. The possible four conformations of evaluators and formatters. We denote each bead in the conformation by its order, from 1 to 12. In order to propagate the Boolean value, all of the conformations start and end at the same height. Arrows at top and bottom denote respectively the previous and updated evaluations of the corresponding clause, though they are in different formats. The purple arrows from top indicate which conformations the **p**-evaluator takes, depending on whether **F** or **T** is input from the left; hence, the **p**-evaluator never takes the conformation (b).

to v_1 from the left and the evaluation from the top as inputs, and outputs the value of v_1 to the right faithfully and the updated evaluation of whether the clause C_k is still satisfiable or not according to the value of v_1 to the bottom. Therefore, four distinct conformations are sufficient for evaluators, and we chose those in Fig. 8. There are three possible ways to update, depending on if C_k includes v_1, or its negation, or none of them. Hence, the OS employs three types of evaluators: **p**, **n**, and **e**. For example, as shown in Fig. 8, the **p**-evaluator folds into the conformation (a) no matter how the clause has been evaluated so far if $v_1 = \mathbf{F}$, while it folds into (c) or (d) depending on the evaluation if $v_1 = \mathbf{T}$. Hence, the **p**-evaluator never folds into (b). The clauses C_1, C_2 and C_3 of the formula evaluated in Fig. 6 include $\neg v_1$, v_1, and none of them, respectively, and hence the first three evaluators are of type **n**, **p**, and **e**, respectively. Note that evaluators output each evaluation in two distinct formats ($\mathbf{U}_1, \mathbf{U}_2$ for unsatisfied, $*_1, *_2$ for satisfiable). They will be reformatted by a submodule called the *formatter* in the succeeding zag, as 10-9-8-7 for $*$ and 10-9-4-3 for $\mathbf{U}$. Analogously, for any $2 \leq i \leq m$, the module w_i first assigns $\mathbf{T}$ or $\mathbf{F}$ randomly to v_i and update the evaluation of clauses provided by the previous zag (of w_{i-1}). As such, the folding of w_i corresponds to the i-th iteration of the for-loop at line 2–6 of Algorithm 1.

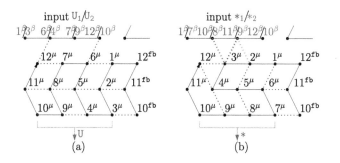

Fig. 9. (a) The glider-like conformation that a formatter takes on the input U_1/U_2. (b) The alternative conformation of a formatter on the input $*_1/*_2$. Blue and red interactions are for the corresponding colored bead only. (Color figure online)

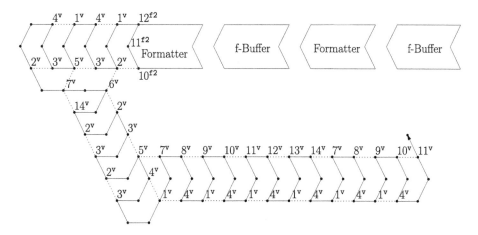

Fig. 10. The first part of the verifier uniquely folds thus and provides a scaffold on which the rest of the verifier serves its role to verify the clauses.

The final module w_v verifies if there is a clause still evaluated to be satisfiable, or equivalently, if the last zag (of w_m) exposes below the sequence 10^{f2}-9^{f2}-8^{f2}-7^{f2}, corresponding to the termination process at line 7–9 (Note that we use superscripts to denote different sets of beads for different types of submodules.). The module is named the *verifier* after this role. The verifier first folds into the conformation shown in Fig. 10. Like the starter, this conformation is also hardcoded; all the rules used are sufficient for the conformation to be folded as shown in Fig. 10. The rest of the verifier is to thread its way from right to left through the recess between the last zag and the floor just made by the first part of the verifier, as shown in Fig. 6. More precisely, it is stretched straight along the floor and once it "detects" a satisfiable clause, or the encoding of $*$, i.e., 10^{f2}-9^{f2}-8^{f2}-7^{f2}, it is pulled up and starts being stretched straight along the zag above. The detection is done by the segment 15^v-16^v-17^v, which we name the *probe*. The probe forms two

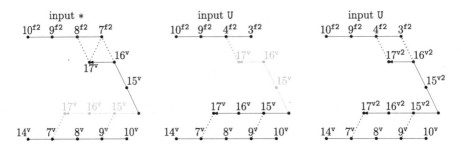

Fig. 11. Detection of satisfiable clauses by the probe segment 15^v-16^v-17^v. (Left) The sequence 8^{f2}-7^{f2}, a part of the encoding of $*$, pulls the probe strongly by 3 h-interactions, one more than the number of h-interactions between the probe and the floor. (Middle) The sequence 4^{f2}-3^{f2}, a part of the encoding of U, cannot pull the probe as strongly as the floor does. (Right) The rule $(3^{f2}, 16^{v2})$ added to $\Xi_{\tau 1}$ allows the probe for C_1 to be pulled upward as well as leftward equally strongly when C_1 is evaluated to be unsatisfied.

h-interactions with the floor, but more h-interactions with the encoding of $*$ due to the rules $(17^v, 8^{f2})$, $(17^v, 7^{f2})$, and $(16^v, 7^{f2})$ (see Fig. 11 (Left)). In contrast, the probe can form only 1 h-interaction with the encoding of U, as shown in Fig. 11 (Middle) due to lack of rules. As a result, the last bead of the probe is stabilized close to the zag above (at the point p_{sat} in Fig. 6) if and only if ϕ is satisfied by the chosen assignment. The last probe is of distinct bead types as 15^{v2}-16^{v2}-17^{v2} for the sake of proving hardness of OSEQ later.

It now suffices to explain the module w_i for the i-th zigzag $(1 \leq i \leq m)$ into detail. Its primary structure is made up as $w_a u_{i,1} \triangleright u_{i,2} \triangleright \cdots \triangleright u_{i,n} \triangleright w_t \triangleleft f_{i,n} \triangleleft \cdots \triangleleft f_{i,2} \triangleleft f_{i,1}$, where w_a is the assignor, w_t is a submodule called the *turner*, and $u_{i,k}$ and $f_{i,k}$ are the evaluator and formatter for the k-th clause, respectively, and triangles ($\triangleright$ and $\triangleleft$) indicate sequences of 12 beads called *e-buffers* and *f-buffers* respectively. Buffers keep a sufficient distance between the consecutive submodules horizontally so as for them not to interact with each other. As shown in Fig. 6, buffers in a consecutive zig and zag may get adjacent to each other vertically. Should they involve a common bead type, an inter-buffer interaction could prevent them from folding into a glider. Therefore, e-buffers and f-buffers use pairwise-distinct sets of bead types. The turner w_t is a hardcoded submodule. Its two possible conformations, shown in Fig. 12, let multiple possible paths arisen from the nondeterministic value assignment to v_i converge into one path (at this point, the system is allowed to "forget" the value).

Evaluator and formatter. The k-th evaluator $u_{i,k}$ and the k-th formatter $f_{i,k}$ in the i-th zig and zag cooperatively update the evaluation of whether the k-th clause is still satisfiable or already unsatisfied. The role of formatters is auxiliary: as we already explained, the output of evaluators ($*$/U) is encoded (as a sequence of beads exposed below) redundantly, and formatters reformat them and ensure that evaluators in the next zig suffice to be capable of reading one sequence of

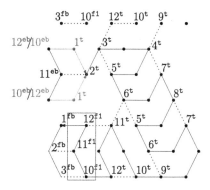

Fig. 12. The two possible conformations of turners, which converge multiple possible paths due to nondeterministic value assignment into one path. Purple box shows the context for the next f-buffer. (Color figure online)

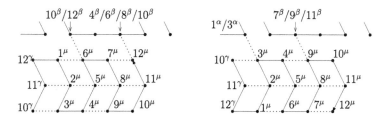

Fig. 13. The basic module is a modification of the glider and folds its primary structure $1^\mu 2^\mu \cdots 12^\mu$ into these two conformations deterministically depending on whether the first bead 1^μ is up or down.

beads for $*$ and one for U. The glider-based foundation, which we will explain shortly, is modified in such a way that the evaluator and formatter inherit the information transfer capability, which enables the n evaluators for the i-th zig, that is, $u_{i,1}, u_{i,2}, \ldots, u_{i,n}$, to transfer the value randomly assigned to v_i one after another.

The basic module is to fold its primary structure $1^\mu 2^\mu \cdots 11^\mu 12^\mu$ (we use Greek letters to represent a set of different bead types) into one of the two gliders shown in Fig. 13 deterministically depending on the two possible contexts (in another context, it could admit another conformation, but in the proposed NOS, the such context is never encountered). It is implemented using the following ruleset R:
$R = \{(2^\mu, 11^\gamma), (3^\mu, 1^\alpha), (3^\mu, 3^\alpha), (3^\mu, 10^\gamma), (5^\mu, 2^\mu), (6^\mu, 11^\gamma), (6^\mu, 1^\mu), (6^\mu, 10^\beta),$
$(6^\mu, 12^\beta), (7^\mu, 10^\gamma), (8^\mu, 5^\mu), (9^\mu, 2^\mu), (9^\mu, 4^\mu), (9^\mu, 7^\beta), (9^\mu, 9^\beta), (9^\mu, 11^\beta), (10^\mu, 1^\mu),$
$(11^\mu, 8^\mu), (12^\mu, 4^\beta), (12^\mu, 6^\beta), (12^\mu, 8^\beta), (12^\mu, 10^\beta), (12^\mu, 3^\mu), (12^\mu, 4^\mu), (12^\mu, 7^\mu)\}$
where $\{(\alpha, \beta, \gamma, \mu)\} = \{(\texttt{fb}, \texttt{f2}, \texttt{eb}, \{\texttt{p1,n1,e1}\}), (\texttt{eb}, \{\texttt{p1,n1,e1}\}, \texttt{fb}, \texttt{f1}), (\texttt{fb}, \texttt{f1}, \texttt{eb},$
$\{\texttt{p2,n2,e2}\}), (\texttt{eb}, \{\texttt{p2,n2,e2}\}, \texttt{fb}, \texttt{f2})\}$.

The evaluator and formatter are derived by "equipping" the basic module with the capability of "reading" the output of the module above; formally speaking, we add some rules to the basic ruleset R that attract some factor of the primary structure (called *input reader*) towards the output so that the resulting module favors another conformation over the glider. Here, one design criterion should be noted: we designed the NOS in such a manner that a module (evaluator/formatter) taking a glider represents the evaluation U. In Fig. 8, the two non-glider conformations are illustrated. Note that these conformations properly propagate the value (F/T) randomly assigned to v_i by the assignor.

Let us focus attention to the evaluator $u_{i,k}$, which evaluates the k-th clause C_k according to the value randomly assigned to v_i and the evaluation made so far by the previous evaluators $u_{1,k}, u_{2,k}, \ldots, u_{i-1,k}$. There are three possibilities to be taken into account depending on whether C_k contains the positive literal v_i, its negation $\neg v_i$, or none of them. That is, three types of evaluators (p, n, and e) are needed. The p-type evaluator is supposed to fold into the glider (U) no matter what the previous evaluation is if $v_i = F$, but be capable of taking both the glider and a non-glider conformation so as to propagate the previous evaluation as it is when $v_i = T$, corresponding to line 5. The n-type evaluator is supposed to behave analogously but the roles of F and T are flipped, corresponding to line 6. The e-type evaluator should propagate the previous evaluation as it is no matter which value is assigned to v_i.

The evaluator $u_{i,k}$ is sandwiched from above and below by two formatters. Since the evaluator interacts with both of them, a bead type common in these formatters would cause misfolding. Therefore, the NOS implements evaluators in consecutive zigs using two pairwise-distinct sets of bead types, even if they are of the identical type. This results in, for instance, two sets of bead types $\{p1, p2\}$ for the type-p evaluators. Similarly, the NOS uses two distinct sets of bead types for each type of evaluators and the formatter, which we distinguish with numbers.

We propose the following two sets R_T, R_F of extra rules, which enable the module to read the output when $v_i = T$ and when $v_i = F$, respectively: $R_T = \{(2^\mu, 9^\beta), (3^\mu, 8^\beta), (4^\mu, 7^\beta), (5^\mu, 10^{fb})\}$ where $\{(\beta, \mu)\} = \{(f2, p1), (f2, e1), (f1, p2), (f1, e2)\}$, and $R_F = \{(5^\mu, 8^\beta), (6^\mu, 7^\beta), (7^\mu, 10^{fb})\}$ where $\{(\beta, \mu)\} = \{(f2, n1), (f2, e1), (f1, n2), (f1, e2)\}$.

Rules in R_T convert the module with sets of bead types $\{p1, p2\}$ into the type-p evaluator, while rules in R_F convert the module with $\{n1, n2\}$ into the type-n evaluator. Note that these extra rulesets do not interfere with each other, and adding both of them converts the module with $\{e1, e2\}$ into the type-e evaluator. Figure 14 shows foldings with newly added rules.

The outputs of an evaluator are redundant: $*$ is encoded as $1^\mu 10^\mu 11^\mu 12^\mu$ or $7^\mu 8^\mu 9^\mu 10^\mu$ whereas U is encoded as $1^\mu 6^\mu 7^\mu 12^\mu$ or $3^\mu 4^\mu 9^\mu 10^\mu$. The redundant output is reformatted by formatters in the i-th zag so as for an input to the evaluators in the next (i+1-th) zig to be encoded in a unique format. Unlike the evaluator, formatters do not have to propagate 1-bit information horizontally. The conformation of the turner fixes the first bead of the first formatter $f_{i,n}$ in the i-th zag up. The following ruleset R_{format} converts the module with sets

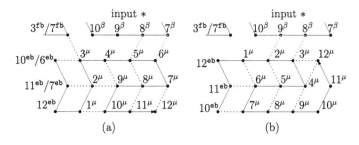

Fig. 14. (a) Folding of n-evaluators and e-evaluators when $v_i = $ F and the input is $*$. Newly added rules are colored in pink. (b) Folding of p-evaluators and e-evaluators when $v_i = $ T and the input is $*$. (Color figure online)

of bead types $\{f1, f2\}$ into the formatter, which takes the conformation (c) in Fig. 8 if the output of the evaluator above is U or the conformation (d) if the output is $*$: $R_{format} = \{(2^\mu, 9^\beta), (2^\mu, 11^\beta), (3^\mu, 8^\beta), (3^\mu, 11^\beta), (4^\mu, 8^\beta), (4^\mu, 1^\beta)\}$ where $\{(\beta, \mu)\} = \{(p1, f1), (n1, f1), (e1, f1), (p2, f2), (n2, f2), (e2, f2)\}$. Figure 9 shows foldings with newly added rules.

We establish the following theorem for Ξ_τ.

Theorem 3. *Let Ξ_τ be the seedless NOS generated from a DNF formula ϕ. Then, ϕ is tautology if and only if there is no conformation of Ξ_τ that reaches the point p_{unsat}.*

Once we establish a robust design of Ξ_τ, we now prove the hardness of OS equivalence test by variations of Ξ_τ. Note that the size of the ruleset in Ξ_τ is constant, and it takes $O(nm)$ time to construct the primary structure of Ξ_τ from a DNF formula with n clauses and m variables. Thus, we can construct Ξ_τ that represents the given formula in $O(nm)$ time. The OSEQ problem admits as a polynomial no-certificate a conformation that one of the two given OSs can fold but the other cannot. It hence belongs to coNP. The coNP-hardness of OSEQ is proved by reduction from DNF tautology. We derive $\Xi_{\tau 1}$ from Ξ_τ by adding one additional rule $(16^{v2}, 3^{f1})$, which makes the verification of $\Xi_{\tau 1}$ nondeterministic when ϕ is not tautology as illustrated in Fig. 11. Thus, ϕ is tautology if and only if Ξ_τ and $\Xi_{\tau 1}$ are equivalent. This proves that OSEQ is coNP-complete even if two OSs are seedless and identical except for their rulesets $\mathcal{H}_1, \mathcal{H}_2$ such that $\mathcal{H}_1 \subseteq \mathcal{H}_2$ and $|\mathcal{H}_2| - |\mathcal{H}_1| = 1$.

4   Conclusions

We have designed a seedless NOS that solves the DNF tautology problem, and demonstrated the hardness of testing the equivalence of two OSs using the designed NOS. Since this is the first attempt to exploit nondeterminism and seedlessness in the design of OS, our future work includes applying nondeterminism and seedlessness to solve other problems. It is an open problem whether we can design an equivalent seedless OS from a given OS or not. Also note that we map an instance of DNF formulas to an NOS that is unique to that input. This notion is called semi-uniformity [8] compared to circuit uniformity [1], where we provide a computing device solely according to the length of the input. Introducing circuit uniformity to the design of OS is another open problem.

Acknowledgments. We would like to thank the anonymous reviewers for the careful reading of the paper and many valuable suggestions.

Kim was supported by NRF Grant funded by the Korean Government (NRF-2013-Global Ph.D. Fellowship Program). The work of S. S. was supported in part by JST Program to Disseminate Tenure Tracking System, MEXT, Japan, No. 6F36, and by JSPS Grant-in-Aid for Research Activity Start-up No. 15H06212 and for Young Scientists (A) No. 16H05854.

References

1. Borodin, A.: On relating time and space to size and depth. SIAM J. Comput. **6**(4), 733–744 (1977)
2. Frieda, K.L., Block, S.M.: Direct observations of cotranscriptional folding in an adenine riboswitch. Science **338**(6105), 397–400 (2012)
3. Geary, C., Meunier, P., Schabanel, N., Seki, S.: Efficient universal computation by greedy molecular folding (2015). CoRR, abs/1508.00510
4. Geary, C., Meunier, P., Schabanel, N., Seki, S.: Programming biomolecules that fold greedily during transcription. In: Proceedings of the 41st International Symposium on Mathematical Foundations of Computer Science (2016, to appear)
5. Geary, C., Rothemund, P.W.K., Andersen, E.S.: A single-stranded architecture for cotranscriptional folding of RNA nanostructures. Science **345**, 799–804 (2014)
6. Kleinberg, J., Tardos, É.: Algorithm Design. Addison-Wesley, Reading (2011)
7. Lai, D., Proctor, J.R., Meyer, I.M.: On the importance of cotranscriptional RNA structure formation. RNA **19**, 1461–1473 (2013)
8. Pérez-Jiménez, M.J., Romero-Jiménez, A., Sancho-Caparrini, F.: Complexity classes in models of cellular computing with membranes. Nat. Comput. **2**(3), 265–285 (2003)
9. Rivas, E., Eddy, S.R.: A dynamic programming algorithm for RNA structure prediction including pseudoknots. J. Mol. Biol. **285**(5), 2053–2068 (1999)
10. Xayaphoummine, A., Bucher, T., Isambert, H.: Kinefold web server for RNA/DNA folding path and structure prediction including pseudoknots and knots. Nucleic Acids Res. **33**, W605–W610 (2005)
11. Zuker, M.: Mfold web server for nucleic acid folding and hybridization prediction. Nucleic Acids Res. **31**(13), 3406–3415 (2003)
12. Zuker, M., Stiegler, P.: Optimal computer folding of large RNA sequences using thermodynamics and auxiliary information. Nucleic Acids Res. **9**(1), 133–148 (1981)

Programming Discrete Distributions with Chemical Reaction Networks

Luca Cardelli[1,2], Marta Kwiatkowska[2], and Luca Laurenti[2(✉)]

[1] Microsoft Research, Cambridge, UK
[2] Department of Computer Science, University of Oxford, Oxford, UK
`luca.laurenti@cs.ox.ac.uk`

Abstract. We explore the range of probabilistic behaviours that can be engineered with Chemical Reaction Networks (CRNs). We show that at steady state CRNs are able to "program" any distribution with finite support in $\mathbb{N}^m$, with $m \geq 1$. Moreover, any distribution with countable infinite support can be approximated with arbitrarily small error under the L^1 norm. We also give optimized schemes for special distributions, including the uniform distribution. Finally, we formulate a calculus to compute on distributions that is complete for finite support distributions, and can be compiled to a restricted class of CRNs that at steady state realize those distributions.

1 Introduction

Individual cells and viruses operate in a noisy environment and molecular interactions are inherently stochastic. How cells can tolerate and take advantage of noise (stochastic fluctuations) is a question of primary importance. It has been shown that noise has a functional role in cells [11]; indeed, some critical functions depend on the stochastic fluctuations of molecular populations and would be impossible in a deterministic setting. For instance, noise is fundamental for probabilistic differentiation of strategies in organisms, and is a key factor for evolution and adaptation [5]. In Escherichia coli, randomly and independently of external inputs, a small sub-population of cells enters a non-growing state in which they can elude the action of antibiotics that can only kill actively growing bacterial cells. Thus, when a population of E. coli cells is treated with antibiotics, the persisted cells survive by virtue of their quiescence before resuming growth [14]. This is an example in which molecular systems compute by producing a distribution. In other cases cells need to shape noise and compute on distributions instead of simply mean values. For example, in [16] the authors show, both mathematically and experimentally, that microRNA confers precision on the protein expression: it shapes the noise of genes in a way that decreases the intrinsic noise in protein expression, maintaining its expected value almost constant. Thus, although fundamentally important, the mechanisms used by cells to compute in a stochastic environment are not well understood.

This research is supported by a Royal Society Research Professorship and ERC AdG VERIWARE.

Y. Rondelez and D. Woods (Eds.): DNA 2016, LNCS 9818, pp. 35–51, 2016.
DOI: 10.1007/978-3-319-43994-5_3

Chemical Reaction Networks (CRNs) with mass action kinetics are a well studied formalism for modelling biochemical systems, more recently also used as a formal programming language [10]. It has been shown that any CRN can be physically implemented by a corresponding DNA strand displacement circuit in a well-mixed solution [18]. DNA-based circuits thus have the potential to operate inside cells and control their activity. Winfree and Qian have also shown that CRNs can be implemented on the surface of a DNA nanostructure [15], enabling localized computation and engineering biochemical systems where the molecular interactions occur between few components. When the number of interacting entities is small, the stochastic fluctuations intrinsic in molecular interactions play a predominant role in the time evolution of the system. As a consequence, "programming" a CRN to provide a particular probabilistic response for a subset of species, for example in response to environmental conditions, is important for engineering complex biochemical nano-devices and randomized algorithms. In this paper, we explore the capacity of CRNs to "program" discrete probability distributions. We aim to characterize the probabilistic behaviour that can be obtained, exploring both the capabilities of CRNs for producing distributions and for computing on distributions by composing them.

Contributions. We show that at steady state CRNs are able to compute any distribution with support in $\mathbb{N}^m$, with $m \geq 1$. We propose an algorithm to systematically "program" a CRN so that its stochastic semantics at steady state approximates a given distribution with arbitrarily small error under the L^1 norm. The approximation is exact if the support of the distribution is finite. The resulting network has a number of reactions linear in the dimension of the support of the distribution and the output is produced monotonically allowing composition. Since distributions with large support can result in unwieldy networks, we also give optimised networks for special distributions, including a novel scheme for the uniform distribution. We formulate a calculus that is complete for finite support distributions, which can be compiled to a restricted class of CRNs that at steady state compute those distributions. The calculus allows for modelling of external influences on the species. Our results are of interest for a variety of scenarios in systems and synthetic biology. For example, they can be used to program a biased stochastic coin or a uniform distribution, thus enabling implementation of randomized algorithms and protocols in CRNs.

Related Work. It has been shown that CRNs with stochastic semantics are Turing complete, up to an arbitrarily small error [17]. If we assume error-free computation, their computational power decreases: they can decide the class of the semi-linear predicates [4] and compute semi-linear functions [9]. A first attempt to model distributions with CRNs can be found in [12], where the problem of producing a single distribution is studied. However, their circuits are approximated and cannot be composed to compute operations on distributions.

2   Chemical Reaction Networks

A *chemical reaction network (CRN)* (Λ, R) is a pair of finite sets, where Λ is the set of *chemical species*, $|\Lambda|$ denotes its size, and R is a set of reactions. A *reaction* $\tau \in R$ is a triple $\tau = (r_\tau, p_\tau, k_\tau)$, where $r_\tau \in \mathbb{N}^{|\Lambda|}$ is the *source complex*, $p_\tau \in \mathbb{N}^{|\Lambda|}$ is the *product complex* and $k_\tau \in \mathbb{R}_{>0}$ is the coefficient associated to the rate of the reaction, where we assume $k_\tau = 1$ if not specified; r_τ and p_τ represent the stoichiometry of reactants and products. Given a reaction $\tau_1 = ([1, 0, 1], [0, 2, 0], k_1)$ we often refer to it as $\tau_1 : \lambda_1 + \lambda_3 \to^{k_1} 2\lambda_2$. The *net change* associated to τ is defined by $v_\tau = p_\tau - r_\tau$.

We assume that the system is well stirred, that is, the probability of the next reaction occurring between two molecules is independent of the location of those molecules, at fixed volume V and temperature. Under these assumptions a *configuration* or *state* of the system $x \in \mathbb{N}^{|\Lambda|}$ is given by the number of molecules of each species. A *chemical reaction system* (CRS) $C = (\Lambda, R, x_0)$ is a tuple where (Λ, R) is a CRN and $x_0 \in \mathbb{N}^{|\Lambda|}$ represents its initial condition.

Stochastic Semantics. In this paper we consider CRNs with stochastic semantics. The propensity rate α_τ of a reaction τ is a function of the current configuration of the system x such that $\alpha_\tau(x)dt$ is the probability that a reaction event occurs in the next infinitesimal interval dt. We assume mass action kinetics [2]. That is, if $\tau : \lambda_1 + ... + \lambda_k \to^k \cdot$, then $\alpha_\tau(x) = k \cdot \prod_{i=1}^{|\Lambda|} \frac{x(\lambda_i)!}{(x(\lambda_i) - r_{\tau,i})!}$, where $r_{\tau,i}$ is the i-th component of vector r.[1] The time evolution of a CRS $C = (\Lambda, R, x_0)$ can be modelled as a time-homogeneous *Continuous Time Markov Chain* (CTMC) $(X^C(t), t \in \mathbb{R}_{\geq 0})$, with state space S [2]. When clear from the context we write $X(t)$ instead of $X^C(t)$. $Q : S \times S \to \mathbb{R}$ is the generator matrix of X, given by $Q(x_i, x_j) = \sum_{\{\tau \in R | x_j = x_i + v_\tau\}} \alpha_\tau(x_i)$ for $i \neq j$ and $Q(x_i, x_i) = -\sum_{j=1 \wedge j \neq i}^{|S|} Q(x_i, x_j)$. We denote $P^C(t)(x) = Prob(X^C(t) = x | X^C(0) = x_0)$, where x_0 is the initial configuration. $P^C(t)(x)$ represents the transient evolution of X, and can be calculated exactly by solving the *Chemical Master Equation (CME)* or by approximation of the CME [7].

Definition 1. *The steady state distribution (or limit distribution) of X^C is defined as $\pi^C = \lim_{t \to \infty} P^C(t)$.*

Again, when clear from the context, instead of π^C we simply write π. π calculates the percentage of time, in the long-run, that X spends in each state $x \in S$. If S is finite, then the above limit distribution always exists and is unique [13]. In this paper we focus on discrete distributions, and will sometimes conflate the term distribution with probability mass function, defined next.

Definition 2. *Suppose that $M : S \to \mathbb{R}^m$ with $m > 0$ is a discrete random variable defined on a countable sample space S. Then the probability mass function (pmf) $f : \mathbb{R}^m \to [0, 1]$ for M is defined as $f(x) = Prob(s \in S | M(s) = x)$.*

---

[1] The reaction rate k depends on the volume V. However, as the volume is fixed, in our notation V is embedded inside k.

For a pmf $\pi : \mathbb{N}^m \to [0,1]$ we call $J = \{y \in \mathbb{N}^m | \pi(y) \neq 0\}$ the support of π. A pmf is always associated to a discrete random variable whose distribution is described by the pmf. However, sometimes, when we refer to a pmf, we imply the associated random variable. Given two pmfs f_1 and f_2 with values in $\mathbb{N}^m$, $m > 0$, we define the L^1 norm (or distance) between them as $d_1(f_1, f_2) = \sum_{n \in \mathbb{N}^m} (|f_1(n) - f_2(n)|)$. Note that, as f_1, f_2 are pmfs, then $d_1(f_1, f_2) \leq 2$. It is worth stressing that, given the CTMC X, for each $t \in \mathbb{R}_{\geq 0}$, $X(t)$ is a random variable defined on a countable state space. As a consequence, its distribution is given by a pmf. Likewise, the limit distribution of a CTMC, if it exists, is a pmf.

Definition 3. *Given* $C = (\Lambda, R)$ *and* $\lambda \in \Lambda$, *we define* $\pi_\lambda(k) = \sum_{\{x \in S | x(\lambda) = k\}} \pi(x)$ *as the probability that for* $t \to \infty$, *in* X^C, *there are* k *molecules of* λ.

π_λ is a pmf representing the steady state distribution of species λ.

3   On Approximating Discrete Distributions with CRNs

We now show that, ffor a pmf with support in $\mathbb{N}$, we can always build a CRS such that, at steady state (i.e. for $t \to \infty$) the random variable representing the molecular population of a given species in the CRN approximates that distribution with arbitrarily small error under the L^1 norm. The result is then generalised to distributions with domain in $\mathbb{N}^m$, with $m \geq 1$. The approximation is exact in case of finite support.

3.1   Programming pmfs

Definition 4. *Given* $f : \mathbb{N} \to [0,1]$ *with finite support* $J = (z_1, ..., z_{|J|})$ *such that* $\sum_{i=1}^{|J|} f(z_i) = 1$, *we define the CRS* $C_f = (\Lambda, R, x_0)$ *as follows.* C_f *is composed of* $2|J|$ *reactions and* $2|J|+2$ *species. For any* $z_i \in J$ *we have two species* $\lambda_i, \lambda_{i,i} \in \Lambda$ *such that* $x_0(\lambda_i) = z_i$ *and* $x_0(\lambda_{i,i}) = 0$. *Then, we consider a species* $\lambda_z \in \lambda$ *such that* $x_0(\lambda_z) = 1$, *and the species* $\lambda_{out} \in \Lambda$, *which represents the output of the network and such that* $x_0(\lambda_{out}) = 0$. *For every* $z_i \in J$, R *has the following two reactions:* $\tau_{i,1} : \lambda_z \xrightarrow{f(z_i)} \lambda_{i,i}$ *and* $\tau_{i,2} : \lambda_i + \lambda_{i,i} \to \lambda_{out} + \lambda_{i,i}$.

Example 1. Consider the probability mass function $f : \mathbb{N} \to [0,1]$ defined as $f(y) = \{\frac{1}{6}$ if $y = 2; \frac{1}{3}$ if $y = 5; \frac{1}{2}$ if $y = 10; 0$ otherwise$\}$. Let $\Lambda = \{\lambda_1, \lambda_2, \lambda_3, \lambda_z, \lambda_{1,1}, \lambda_{2,2}, \lambda_{3,3}, \lambda_{out}\}$, then we build the CRS $C = (\Lambda, R, x_0)$ following Definition 4, where R is given by the following set of reactions:

$$\lambda_z \xrightarrow{\frac{1}{6}} \lambda_{1,1}; \quad \lambda_z \xrightarrow{\frac{1}{3}} \lambda_{2,2}; \quad \lambda_z \xrightarrow{\frac{1}{2}} \lambda_{3,3};$$

$$\lambda_1 + \lambda_{1,1} \xrightarrow{1} \lambda_{1,1} + \lambda_{out}; \quad \lambda_2 + \lambda_{2,2} \xrightarrow{1} \lambda_{2,2} + \lambda_{out}; \quad \lambda_3 + \lambda_{3,3} \xrightarrow{1} \lambda_{3,3} + \lambda_{out}.$$

The initial condition x_0 is $x_0(\lambda_{out}) = x_0(\lambda_{1,1}) = x_0(\lambda_{2,2}) = x_0(\lambda_{3,3}) = 0$; $x_0(\lambda_1) = 2; x_0(\lambda_2) = 5; x_0(\lambda_3) = 10; x_0(\lambda_z) = 1$. Theorem 1 ensures $\pi_{\lambda_{out}} = f$.

Theorem 1. *Given a pmf* $f : \mathbb{N} \to [0,1]$ *with finite support* J, *the CRS* C_f *as defined in Definition 4 is such that* $\pi_{\lambda_{out}}^{C_f} = f$.

Proof. Let $J = (z_1, .., z_{|J|})$ be the support of f, and $|J|$ its size. Suppose $|J|$ is finite, then the set of reachable states from x_0 is finite by construction and the limit distribution of X^{C_f}, the induced CTMC, exists. By construction, in the initial state x_0 only reactions of type $\tau_{i,1}$ can fire, and the probability that a specific $\tau_{i,1}$ fires first is exactly:

$$\frac{\alpha_{\tau_{i,1}}(x_0)}{\sum_{j=1}^{|J|} \alpha_{\tau_{j,1}}(x_0)} = \frac{f(z_i) \cdot 1}{\sum_{j=1}^{|J|} f(z_j) \cdot 1} = \frac{f(z_i)}{\sum_{j=1}^{|J|} f(z_j)} = \frac{f(z_i)}{1} = f(z_i)$$

Observe that the firing of the first reaction uniquely defines the limit distribution of X^{C_f}, because λ_z is consumed immediately and only reaction $\tau_{i,2}$ can fire, with no race condition, until λ_i are consumed. This implies that at steady state λ_{out} will be equal to $x_0(\lambda_i)$, and this happens with probability $f(x_0(\lambda_i))$. Since $x_0(\lambda_i) = z_i$ for $i \in [1, |J|]$, we have $\pi_{\lambda_{out}}^{C_f} = f$. $\square$

Then, we can state the following corollary of Theorem 1.

Corollary 1. *Given a pmf* $f : \mathbb{N} \to [0,1]$ *with countable support* J, *we can always find a finite CRS* C_f *such that* $\pi_{\lambda_{out}}^{C_f} = f$ *with arbitrarily small error under the* L^1 *norm.*

Proof. Let $J = \{z_1, ..., z_{|J|}\}$. Suppose J is (countably) infinite, that is, $|J| \to \infty$. Then, we can always consider an arbitrarily large but finite number of points in the support, such that the probability mass lost is arbitrarily small, and applying Definition 4 on this finite subset of the support we have the result.

In order to prove the result consider the function f' with support $J' = \{z_1, ..., z_k\}$, $k \in \mathbb{N}$, such that $f(z_i) = f'(z_i)$, for all $i \in \mathbb{N}_{\leq k}$. Consider the series $\sum_{i=1}^{\infty} f(n)$. This is an absolute convergent series by definition of pmf. Then, we have that $\lim_{i \to \infty} f(i) = 0$ and, for any $\epsilon > 0$, we can choose some $\kappa_\epsilon \in \mathbb{N}$, such that:

$$\forall k > \kappa_\epsilon \quad |\sum_{i=1}^{k} f'(i) - \sum_{i=1}^{\infty} f(i)| < \frac{\epsilon}{2}.$$

This implies that for $k > \kappa_\epsilon$ given $f_k' = \sum_{i=1}^{k} f'(i)$ we have, $d_1(f_k', f) < \epsilon$. $\square$

The following remark shows that the need for precisely tuning the value of reaction rates in Theorem 1 can be dropped by introducing some auxiliary species.

Remark 1. In practice, tuning the rates of a reaction can be difficult or impossible. However, it is possible to modify the CRS derived using Definition 4 in such a way the probability value is not encoded in the rates, and we just require that all reactions have the same rates. We can do that by using some auxiliary species $\Lambda_c = \{\lambda_{c_1}, \lambda_{c_2}, ..., \lambda_{c_{|\Lambda_c|}}\}$. Then, the reactions $\tau_{i,1}$ for $i \in [1, J]$ become $\tau_{i,1} : \lambda_z + \lambda_{c_i} \rightarrow^k \lambda_{i,i}$, for $k \geq 0$, instead of $\tau_{i,1} : \lambda_z \rightarrow^{f(y_i)} \lambda_{i,i}$, as in the original definition. The initial condition of λ_{c_i} is $x_0(\lambda_{c_i}) = f(y_i) \cdot L$, where $L \in \mathbb{N}$ is such that for $j \in [1, |J|]$ and $J = \{z_1, ..., z_{|J|}\}$ we have that $f(z_j) \cdot L$ is a natural number, assuming all the $f(z_j)$ are rationals.

Remark 2. In biological circuits the probability distribution of a species may depend on some external conditions. For example, the *lambda Bacteriofage* decides to lyse or not to lyse with a probabilistic distribution based also on environmental conditions [5]. Programming similar behaviour is possible by extension of Theorem 1. For instance, suppose, we want to program a switch that with rate $50 + Com$ goes in a state O_1, and with rate 5000 goes in a different state O_2, where Com is an external input. To program this logic we can use the following reactions: $\tau_{1,1} : \lambda_z + \lambda_{c_1} \rightarrow^{k_1} \lambda_{O_1}$ and $\tau_{1,2} : \lambda_z + \lambda_{c_2} \rightarrow^{k_1} \lambda_{O_2}$, where λ_{O_1} and λ_{O_2} model the two logic states, initialized at 0. The initial condition x_0 is such that $x_0(\lambda_z) = 1$, $x_0(\lambda_{c_1}) = 50$ and $x_0(\lambda_{c_2}) = 5000$. Then, we add the following reaction $Com \rightarrow^{k_2} \lambda_{c_1}$. It is easy to show that if $k_2 >>> k_1$ then we have the desired probabilistic behaviour for any initial value of $Com \in \mathbb{N}$. This may be of interest also for practical scenarios in synthetic biology, where for instance the behaviour of synthetic bacteria needs to be externally controlled [3]; and, if each bacteria is endowed with a similar logic, then, by tuning Com, at the population level, it is possible to control the fraction of bacteria that perform this task.

In the next theorem we generalize to the multi-dimensional case.

Theorem 2. *Given $f : \mathbb{N}^m \rightarrow [0, 1]$ with $m \geq 1$ such that $\sum_{i \in \mathbb{N}} f(i) = 1$, then there exists a CRS $C = (\Lambda, R, x_0)$ such that the joint limit distribution of $(\lambda_{out_1}, \lambda_{out_2}, ..., \lambda_{out_m}) \in \Lambda$ approximates f with arbitrarily small error under the L^1 distance. The approximation is exact if the support of f is finite.*

To prove this theorem we can derive a CRS similar to that in the uni-dimensional case. The firing of the first reaction can be used to probabilistically determine the value at steady state of the m output species, using some auxiliary species.

3.2   Special Distributions

For a given pmf the number of reactions of the CRS derived from Definition 4 is linear in the dimension of its support. As a consequence, if the support is large then the CRSs derived using Theorems 1 and 2 can be unwieldy. In the following we show three optimised CRSs to calculate the Poisson, binomial and uniform distributions. These CRNs are compact and applicable in many practical scenarios. However, using Definition 4 the output is always produced monotonically. In the circuits below this does not happen, but, on the other hand, the gain in compactness is substantial. The first two circuits have been derived from the literature, while the CRN for the uniform distribution is new.

Poisson Distribution. The main result of [1] guarantees that all the CRNs that respect some conditions (weakly reversible, deficiency zero and irreducible state space, see [1]) have a distribution given by the product of Poisson distributions. As a particular case, we consider the following CRS composed of only one species λ and the following two reactions $\tau_1 : \emptyset \to^{k_1} \lambda; \tau_2 : \lambda \to^{k_2} \emptyset$. Then, at steady state, λ has a Poisson distribution with expected value $\frac{k_1}{k_2}$.

Binomial Distribution. We consider the network introduced in [1]. The CRS is composed of two species, λ_1 and λ_2, with initial condition x_0 such that $x_0(\lambda_1) + x_0(\lambda_2) = K$ and the following set of reactions: $\tau_1 : \lambda_1 \to^{k_1} \lambda_2; \tau_2 : \lambda_2 \to^{k_2} \lambda_1$. As shown in [1], λ_1 and λ_2 at steady state have a binomial distribution such that: $\pi_{\lambda_1}(y) = \binom{K}{y} c_1{}^y (1 - c_1)^{K-y}$ and $\pi_{\lambda_2}(y) = \binom{K}{y} c_2{}^y (1 - c_2)^{K-y}$.

Uniform Distribution. The following CRS computes the uniform distribution over the sum of the initial number of molecules in the system, independently of the initial value of each species. It has species λ_1 and λ_2 and reactions:

$$\tau_1 : \lambda_1 \to^k \lambda_2; \quad \tau_2 : \lambda_2 \to^k \lambda_1; \quad \tau_3 : \lambda_1 + \lambda_2 \to^k \lambda_1 + \lambda_1; \quad \tau_4 : \lambda_1 + \lambda_2 \to^k \lambda_2 + \lambda_2$$

For $k > 0$, τ_1 and τ_2 implement the binomial distribution. These are combined with τ_3 and τ_4, which implement a Direct Competition system [6], which has a bimodal limit distribution in 0 and in K, where $x_0(\lambda_1) + x_0(\lambda_2) = K$, with x_0 initial condition. This network, surprisingly, according to the next theorem, at steady state produces a distribution which varies uniformly between 0 and K.

Theorem 3. *Let $x_0(\lambda_1) + x_0(\lambda_2) = K \in \mathbb{N}$. Then, the CRS described above has the following steady state distribution for λ_1 and λ_2:*

$$\pi_{\lambda_1}(y) = \pi_{\lambda_2}(y) = \begin{cases} \frac{1}{K+1}, & \text{if } y \in [0, K] \\ 0, & \text{otherwise} \end{cases}$$

Proof. We consider a general initial condition x_0 such that $x_0(\lambda_1) = K - M$ and $x_0(\lambda_2) = M$ for $0 \leq M \leq K$ and $K, M \in \mathbb{N}$. Because any reaction has exactly 2 reagents and 2 products, we have the invariant that for any configuration x reachable from x_0 it holds that $x(\lambda_1) + x(\lambda_2) = K$. Figure 1 plots the CTMC semantics of the system. For any fixed K the set of reachable states from any initial condition in the induced CTMC is finite (exactly K states are reachable from any initial condition) and irreducible. Therefore, the steady state solution exists, is unique and independent of the initial conditions. To find this limit distribution we can calculate Q, the infinitesimal generator of the CTMC, and then solve the linear equations system $\pi Q = 0$, with the constraint that $\sum_{i \in [0,K]} \pi_i = 1$, where π_i is the ith component of the vector π, as shown in [13]. Because the CTMC we are considering is irreducible, this is equivalent to solve the balance equations with the same constraint. The resulting π is the steady state distribution of the system.

We consider 3 cases, where $(K - j, j)$ for $j \in [0, K]$ represents the state of the system in terms of molecules of λ_1 and λ_2.

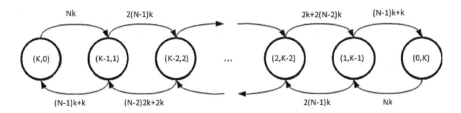

Fig. 1. The figure shows the CTMC induced by the CRS implementing the uniform distribution for initial condition x_0 such that $x_0(\lambda_1) + x_0(\lambda_2) = K$.

- Case $j = 0$. For the state $(K, 0)$, whose limit distribution is defined as $\pi(K, 0)$, we have the following balance equation:

$$-\pi(K, 0)Kk + \pi(K - 1, 1)[(K - 1)k + k] = 0 \implies$$
$$\pi(K, 0) = \pi(K - 1, 1).$$

- Case $j \in [1, K - 1]$. Observing Fig. 1 we see that the states and the rates follow a precise pattern: every state is directly connected with only two states and for any transition the rates depend on two reactions, therefore we can consider the balance equations for a general state $(K - j, j)$ for $j \in [1, K - 1]$ (for the sake of a lighter notation instead of $\pi(K - j, j)$ we write π^j):

$$\pi^{j-1}[K + 1 - j + (K + 1 - j)(j - 1)]$$
$$- \pi^j[2(K - j)j + j + K - j] + \pi^{j+1}[j + 1 + (K - j - 1)(j + 1)] = 0$$
$$\implies$$
$$\pi^{j-1}[Kj - j^2 + j] - \pi^j[2Kj - 2j^2 + K] + \pi^{j+1}[Kj + K - j^2 - j] = 0$$

It is easy to verify that if $\pi^{j-1} = \pi^j = \pi^{j+1}$ then the equation is verified.
- Case $j = K$. The case for the state $(0, K)$ is similar to the case $(K, 0)$.

We have shown that each reachable state has equal probability at steady state for any possible initial condition. Therefore, because $\sum_{i=0}^{K} \pi^i = 1$ and $\pi_{\lambda_i}(y) = \sum_{x_j \in S | x_j(\lambda_i) = y} \pi^j$ for $y \geq 0$, we have that for both λ_1 and λ_2

$$\pi_{\lambda_1}(y) = \pi_{\lambda_2}(y) = \begin{cases} \frac{1}{K+1}, & \text{if } y \in [0, K] \\ 0, & \text{otherwise} \end{cases}$$

$\square$

4   Calculus of Limit Distributions of CRNs

In the previous section we have shown that CRNs are able to program any pmf on $\mathbb{N}$. We now define a calculus to compose and compute on pmfs. We show it is complete with respect to finite support pmfs on $\mathbb{N}$. Then, we define a translation of this calculus into a restricted class of CRNs. We prove the soundness of such a translation, which thus yields an abstract calculus of limit distributions of CRNs. For simplicity, in what follows we consider only pmfs with support in $\mathbb{N}$, but the results can be generalised to the multi-dimensional case in a straightforward way.

Definition 5 (*Syntax*). *The syntax of formulae of our calculus is given by*

$$P := (P + P) \mid min(P, P) \mid k \cdot P \mid (P)_D : P \mid one \mid zero$$

$$D := p \mid p \cdot c_i + D$$

where $k \in \mathbb{Q}_{\geq 0}$, $p \in \mathbb{Q}_{[0,1]}$ are rational and $V = \{c_1, ..., c_n\}$ is a set of variables with values in $\mathbb{N}$.

A formula P denotes a pmf that can be obtained as a sum, minimum, multiplication by a rational, or convex combination of pmfs *one* and *zero*. Given a formula P, variables $V = \{c_1, ..., c_n\}$, called *environmental inputs*, model the influence of external factors on the probability distributions of the system. $V(P)$ represents the variables in P. An *environment* $E : V \to \mathbb{Q}_{[0,1]}$ is a partial function which maps each input c_i to its valuation normalized to $[0, 1]$. Given a formula P and an environment E, where $V(P) \subseteq dom(E)$, with $dom(E)$ domain of E, we define its semantics, $[\![P]\!]_E$, as a pmf (the empty environment is denoted as $\emptyset$). D expresses a summation of valuations of inputs c_i weighted by rational probabilities p, which evaluates to a rational $[\![D]\!]_E$ for a given environment. We require that, for any D, the sum of p coefficients in D is in $[0, 1]$. This ensures that $0 \leq [\![D]\!]_E \leq 1$. The semantics is defined inductively as follows, where the operations on pmfs are defined in Sect. 4.1.

Definition 6 (*Semantics*). *Given formulae P, P_1, P_2 and an environment E, such that $V(P) \cup V(P_1) \cup V(P_2) \subseteq dom(E)$, we define*

$$[\![one]\!]_E = \pi_{one} \quad [\![zero]\!]_E = \pi_{zero} \quad [\![P_1 + P_2]\!]_E = [\![P_1]\!]_E + [\![P_2]\!]_E$$

$$[\![min(P_1, P_2)]\!]_E = min([\![P_1]\!]_E, [\![P_2]\!]_E)$$

$$[\![k \cdot P]\!]_E = \frac{k_1 \cdot ([\![P]\!]_E)}{k_2} \quad for \ k = \frac{k_1}{k_2} \ and \ k_1, k_2 \in \mathbb{N}$$

$$[\![(P_1)_D : (P_2)]\!]_E = ([\![P_1]\!]_E)_{[\![D]\!]_E} : ([\![P_2]\!]_E)$$

$$[\![p]\!]_E = p \quad [\![p \cdot c_i + D]\!]_E = p \cdot E(c_i) + ([\![D]\!]_E)$$

$$where \ \pi_{one}(y) = \begin{cases} 1, & if \ y = 1 \\ 0, & otherwise \end{cases} \ and \ \pi_{zero}(y) = \begin{cases} 1, & if \ y = 0 \\ 0, & otherwise \end{cases}.$$

To illustrate the calculus, consider the Bernoulli distribution with parameter $p \in \mathbb{Q}_{[0,1]}$. We have $bern^p = (one)_p : zero$, where $[\![bern^p]\!]_\emptyset(y) = \{p$ if $y = 1; 1 - p$ if $y = 0; 0$ otherwise$\}$. The binomial distribution can be obtained as a sum of n independent Bernoulli distributions of the same parameter. Given a random variable with a binomial distribution with parameters (n, p), if n is sufficiently large and p sufficiently small then this approximates a Poisson distribution with parameter $n \cdot p$.

4.1   Operations on Distributions

In this section, we define a set of operations on pmfs needed to define the semantics of the calculus. We conclude the section by showing that these operations are sufficient to represent pmfs with finite support in $\mathbb{N}$.

Definition 7. *Let $\pi_1 : \mathbb{N} \to [0,1]$, $\pi_2 : \mathbb{N} \to [0,1]$ be two pmfs. Assume $p \in \mathbb{Q}_{[0,1]}$, $y \in \mathbb{N}$, $k_1 \in \mathbb{N}$ and $k_2 \in \mathbb{N}_{>0}$, then we define the following operations on pmfs:*

- *The sum or convolution of π_1 and π_2 is defined as $(\pi_1 + \pi_2)(y) = \sum_{(y_i,y_j) \in \mathbb{N} \times \mathbb{N} \ s.t. \ y_i + y_j = y} \pi_1(y_i)\pi_2(y_j)$.*
- *The minimum of π_1 and π_2 is defined as $min(\pi_1, \pi_2)(y) = \sum_{(y_i,y_j) \in \mathbb{N} \times \mathbb{N} \ s.t. \ min(y_i,y_j) = y} \pi_1(y_i)\pi_2(y_j)$.*
- *The multiplication of π_1 by the constant k_1 is defined as $(k_1 \pi_1)(y) = \begin{cases} \pi_1(\frac{y}{k_1}), & if \ \frac{y}{k_1} \in \mathbb{N} \\ 0, & otherwise \end{cases}$*
- *The division of π_1 by the constant k_2 is defined as $\frac{\pi}{k_2}(y) = \sum_{y_i \in \mathbb{N} \ s.t. \ y = \lfloor y_i / k_2 \rfloor} \pi(y_i)$.*
- *The convex combination of π_1 and π_2, for $y \in \mathbb{N}$, is defined $((\pi_1)_p : (\pi_2))(y) = p\pi_1(y) + (1 - p)\pi_2(y)$.*

The convex combination operator is the only one that is not closed with respect to pmfs whose support is a single point. Lemma 1 shows that this operator is not associative with respect to minimum and sum of pmfs.

Lemma 1. *Given probability mass functions π_1, $\pi_2 : \mathbb{N} \to [0,1]$, $p_1, p_2, p_3, p_4 \in [0,1]$ and $k \in \mathbb{Q}_{\geq 0}$, then the following equations hold:*

- $k((\pi_1)_p : \pi_2) = (k\pi_1)_p : (k\pi_2)$
- $((\pi_1)_{p_1} : \pi_2)_{p_2} : \pi_3 = (\pi_1)_{p_3} : ((\pi_2)_{p_4} : \pi_3)$ *iff $p_3 = p_1 p_2$ and $p_4 = \frac{(1-p_1)p_2}{1-p_1 p_2}$*
- $(\pi_1)_p : \pi_2 = (\pi_2)_{1-p} : \pi_1$
- $(\pi_1)_p : \pi_1 = \pi_1$.

Example 2. Consider the following formula

$$P_1 = (one)_{0.001 \cdot c + 0.2}(4 \cdot one) + (2 \cdot one)_{0.4}(3 \cdot one),$$

with set of environmental variables $V = \{c\}$ and an enviroment E such that $V(P_1) \subseteq dom(E)$. Then, according to Definition 7 we have that

$$[\![P_1]\!]_E(y) = \begin{cases} (0.001 \cdot [\![c]\!]_E + 0.2) \cdot 0.4, & if \ y = 3 \\ (0.001 \cdot [\![c]\!]_E + 0.2) \cdot 0.6, & if \ y = 4 \\ (1 - (0.001 \cdot [\![c]\!]_E + 0.2)) \cdot 0.4, & if \ y = 6 \\ (1 - (0.001 \cdot [\![c]\!]_E + 0.2)) \cdot 0.6, & if \ y = 7 \\ 0, & otherwise \end{cases}$$

Having formally defined all the operations on pmfs, we can finally state the following proposition guaranteeing that the semantics of any formula of the calculus is a pmf.

Proposition 1. *Given P, a formula of the calculus defined in Definition 5, and an environment E such that $V(P) \subseteq dom(E)$, then $[\![P]\!]_E$ is a pmf.*

The following theorem shows that our calculus is complete with respect to finite support distributions.

Theorem 4. *For any pmf $f : \mathbb{N} \to [0,1]$ with finite support there exists a formula P such that $[\![P]\!]_\emptyset = f$.*

Proof. Given a pmf $f : \mathbb{N} \to [0,1]$ with finite support $J = (z_1, ..., z_{|J|})$ we can define $P = (z_1 \cdot one)_{f(z_1)} : ((z_2 \cdot one)_{\frac{f(z_2)}{1-f(z_1)}} : (... : ((z_i \cdot one)_{\frac{f(z_i)}{\prod_{j=1}^{i-1}(1-f(z_j))}} : ... :$
$((z_n \cdot one)))))$. Then, $[\![P]\!]_\emptyset = f$.   □

Proof of Theorem 4 relies only on a subset of the operators, but the other operators are useful for composing previously defined pmfs.

5 CRN Implementation

We now show how the operators of the calculus can be realized by operators on CRSs. The resulting CRSs produce the required distributions at steady state, that is, in terms of the steady state distribution of the induced CTMC. Thus, we need to consider a restricted class of CRNs that always stabilize and that can be incrementally composed. The key idea is that each such CRN has output species that cannot act as a reactant in any reaction, and hence the counts of those species increase monotonically.[2] This implies that the optimized CRSs shown in Sect. 3.2 cannot be used compositionally.

5.1 Non-Reacting Output CRSs (NRO-CRSs)

Since in the calculus presented in Definition 5 we consider only finite support pmfs, in this section we are limited to finite state CTMCs. This is important because some results valid for finite state CTMCs are not valid in infinite state spaces. Moreover, any pmf with infinite support on natural numbers can always be approximated under the L^1 norm (see Corollary 1).

Given a CRS $C = (\Lambda, R, x_0)$, we call the *non-reacting species* of C the subset of species $\Lambda_r \subseteq \Lambda$ such that given $\lambda_r \in \Lambda_r$ there does not exist $\tau \in R$ such that $r_\tau^{\lambda_r} > 0$, where $r_\tau^{\lambda_r}$ is the component of the source complex of the reaction τ relative to λ_r, that is, λ_r is not a reactant in any reaction. Given C we also define a subset of species, $\Lambda_o \subseteq \Lambda$, as the *output species* of C. Output species are those whose limit distribution is of interest. In general, they may or may not be *non-reacting species*; they depend on the observer and on what he/she is interested in observing.

---

[2] Note that this is a stricter requirement than those in [9], where output species are produced monotonically, but they are allowed to act as catalysts in some reactions. We cannot allow that because catalyst species influence the value of the propensity rate of a reaction and so the probability that it fires.

Definition 8. *A non-reacting output CRS (NRO-CRS) is a tuple $C = (\Lambda, \Lambda_o, R, x_0)$, where $\Lambda_o \subseteq \Lambda$ are the output species of C such that $\Lambda_o \subseteq \Lambda_r$, where Λ_r are the non-reacting species of C.*

NRO-CRNs are CRSs in which the output species are produced monotonically and cannot act as a reactant in any reaction. A consequence of Theorem 1 is the following lemma, which shows that this class of CRNs can approximate any pmf with support on natural numbers, up to an arbitrarily small error.

Lemma 2. *For any probability mass function $f : \mathbb{N}^m \to [0,1]$ there exists a NRO-CRS such that the joint limit distribution of its output species approximates f with arbitrarily small error under the L^1 norm. The approximation is exact if the support of f is finite.*

In Table 1, we define a set of operators on NRO-CRSs. Let $C_1 = (\Lambda_1, \Lambda_{o_1}, R_1, x_{0_1})$ and $C_2 = (\Lambda_2, \Lambda_{o_2}, R_2, x_{0_2})$ be NRO-CRSs such that $\Lambda_1 \cap \Lambda_2 = \emptyset$. Let $\lambda_{o_1} \in \Lambda_{o_1}$ and $\lambda_{o_2} \in \Lambda_{o_2}$, then we define the set of reactions which implements the operators of Sum, Minimum, Multiplication by a constant $k_1 \in \mathbb{N}$ and Division by a constant $k_2 \in \mathbb{N}_{\geq 0}$ over the steady state distribution of λ_{o_1} and λ_{o_2}. The output species of each composed NRO-CRS is λ_{out}, and we assume $\{\lambda_{out}\} \cap (\Lambda_1 \cup \Lambda_2) = \emptyset$ and $x_0(\lambda_{out}) = 0$.

Table 1. CRS operators

Operator	Resulting NRO-CRS
Sum	$(\Lambda_1 \cup \Lambda_2 \cup \{\lambda_{out}\}, \{\lambda_{out}\}, R_1 \cup R_2 \cup \{\lambda_{o_1} \to \lambda_{out}, \lambda_{o_2} \to \lambda_{out}\}, x_0)$
Min	$(\Lambda_1 \cup \Lambda_2 \cup \{\lambda_{out}\}, \{\lambda_{out}\}, R_1 \cup R_2 \cup \{\lambda_{o_1} + \lambda_{o_2} \to \lambda_{out}\}, x_0)$
Mul by k_1	$(\Lambda_1 \cup \{\lambda_{out}\}, \{\lambda_{out}\}, R_1 \cup \{\lambda_{o1} \to \underbrace{\lambda_{out} + \dots + \lambda_{out}}_{k_1 \; times}\}, x_0)$
Div by k_2	$(\Lambda_1 \cup \{\lambda_{out}\}, \{\lambda_{out}\}, R_1 \cup \{\underbrace{\lambda_{o_1} + \dots + \lambda_{o_1}}_{k_2 \; times} \to \lambda_{out}\}, x_0)$

We emphasize that proving that CRS operators of Table 1 implement the operations in Definition 7 is not trivial. In fact, we need to compose stochastic processes and show that the resulting process has the required properties. Fundamental to that end is a convenient representation of X in terms of a summation of time-inhomogeneous Poisson processes, one for each reaction [2]. In what follows we present in slightly extended form the operators for convex combination, with or without external inputs (respectively $Con(\cdot)$ and $ConE(\cdot)$). Formal definitions and proofs of correctness of all the circuits are presented in [8].

Considering C_1 and C_2, as previously, then we need to derive a CRS operator $Con(C_1, \lambda_{o_1}, C_2, \lambda_{o_2}, p, \lambda_{out})$ such that $\pi_{\lambda_{out}} = (\pi_{\lambda_{o_1}}^{C_1})_p : (\pi_{\lambda_{o_2}}^{C_2})$. That is, at steady stade, λ_{out} equals $\pi_{\lambda_{o_1}}^{C_1}$ with probability p and $\pi_{\lambda_{o_2}}^{C_2}$ with probability $1 - p$. This can be done by using Theorem 2 to generate a bi-dimensional synthetic coin with output species $\lambda_{r_1}, \lambda_{r_2}$ such that their joint limit distribution is $\pi_{\lambda_{r_1},\lambda_{r_2}}(y_1, y_2) = \{p$ if $y_1 = 1$ and $y_2 = 0; 1 - p$ if $y_1 = 0$ and $y_2 = 1; 0$ otherwise$\}$. That is, λ_{r_1} and λ_{r_2} are mutually exclusive at steady state. Using these species as catalysts in $\tau_3 : \lambda_{o_1} + \lambda_{r_1} \to \lambda_{r_1} + \lambda_{out}$ and $\tau_4 : \lambda_{o_2} + \lambda_{r_2} \to \lambda_{r_2} + \lambda_{out}$ we have exactly the desired result at steady state.

Example 3. Consider the following NRO-CRSs $C_1 = (\{\lambda_{o_1}\}, \{\lambda_{o_1}\}, \{\}, x_{0_1})$ and $C_2 = (\{\lambda_{o_2}\}, \{\lambda_{o_2}\}, \{\}, x_{0_2})$, with initial condition $x_{0_1}(\lambda_{o_1}) = 10$ and $x_{0_2}(\lambda_{o_2}) = 20$. Then, the operator $ConE(C_1, \lambda_{o_1}, C_2, \lambda_{o_2}, 0.3, \lambda_{out})$ implements the operation $\pi_{\lambda_{out}} = (\pi_{\lambda_{o_1}}^{C_1})_{0.3}(\pi_{\lambda_{o_2}}^{C_2})$ and it is given by the following reactions:

$$\lambda_z \xrightarrow{0.3} \lambda_{r_1}; \quad \lambda_z \xrightarrow{0.7} \lambda_{r_2}; \quad \lambda_{r_1} + \lambda_{o_1} \to \lambda_{r_1} + \lambda_{out}; \quad \lambda_{r_2} + \lambda_{o_2} \to \lambda_{r_2} + \lambda_{out}$$

with initial condition x_0 such that $x_0(\lambda_z) = 1, x_0(\lambda_{r_1}) = x_0(\lambda_{r_2}) = x_0(\lambda_{out}) = 0$.

Let C_1, C_2 be as above and $f = p_0 + p_1 \cdot c_1 + ... + p_n \cdot c_n$ with $p_1, ..., p_n \in \mathbb{Q}_{[0,1]}$, $V = \{c_1, ..., c_n\}$ a set of environmental variables, and E, an environment such that $V \subseteq dom(E)$. Then, computing a CRS operator $ConE(C_1, \lambda_{o_1}, C_2, \lambda_{o_2}, f(E(V)), \lambda_{out})$ such that $\pi_{\lambda_{out}} = (\pi_{\lambda_{o_1}}^{C_1})_{f(E(V))} : (\pi_{\lambda_{o_2}}^{C_2})$ is a matter of extending the previous circuit. First of all, we can derive the CRS to compute $f(E(V))$ and $1 - f(E(V))$ and memorize them in some species. This can be done as $f(E(V))$ is semi-linear [9]. Then, as $f(E(V)) \leq 1$ by assumption, we can use these species as catalysts to determine the output value of λ_{out}, as in the previous case. As shown in [8], this circuit, in the case of external inputs, introduces an arbitrarily small, but non-zero, error, due to the fact that there is no way to know when the computation of $f(E(V))$ terminates.

Example 4. Consider the following NRO-CRSs $C_1 = (\{\lambda_{o_1}\}, \{\lambda_{o_1}\}, \{\}, x_{0_1})$ and $C_2 = (\{\lambda_{o_2}\}, \{\lambda_{o_2}\}, \{\}, x_{0_2})$, with initial condition $x_{0_1}(\lambda_{o_1}) = 10$ and $x_{0_2}(\lambda_{o_2}) = 20$. Then, consider the following functions $f(E(c)) = E(c)$, where E is a partial function assigning values to c, and it is assumed $0.001 \leq E(c) \leq 1$ and that $E(c) \cdot 1000 \in \mathbb{N}$. Then, the operator $ConE(C_1, \lambda_{o_1}, C_2, \lambda_{o_2}, f, \lambda_{out})$, implements the operation $\pi_{\lambda_{out}} = (\pi_{\lambda_{o_1}}^{C_1})_{E(c)}(\pi_{\lambda_{o_2}}^{C_2})$ and it is given by the following reactions:

$$\tau_1 : \lambda_c \xrightarrow{k_1} \lambda_{Cat_1} + \lambda_{Cat_2}; \quad \tau_2 : \lambda_{Tot} + \lambda_{Cat_2} \xrightarrow{k_1} \emptyset$$
$$\tau_3 : \lambda_z + \lambda_{Cat_1} \xrightarrow{k_2} \lambda_1; \quad \tau_4 : \lambda_z + \lambda_{Tot} \xrightarrow{k_2} \lambda_2$$
$$\tau_5 : \lambda_{o_1} + \lambda_1 \xrightarrow{k_2} \lambda_1 + \lambda_{out}; \quad \tau_6 : \lambda_{o_2} + \lambda_2 \xrightarrow{k_2} \lambda_2 + \lambda_{out}$$

where $\lambda_c, \lambda_{Cat_1}, \lambda_{Cat_2}, \lambda_z, \lambda_1, \lambda_2$ are auxiliary species with initial condition x_0 such that $x_0(\lambda_{Cat_1}) = x_0(\lambda_{Cat_2}) = x_0(\lambda_1) = x_0(\lambda_2) = 0$, $x_0(\lambda_{Tot}) = 1000$, $x_0(\lambda_z) = 1$, $x_0(\lambda_c) = E(c) \cdot 1000$ and $k_1 \gg k_2$. Reactions τ_1, τ_2 implement $f(E(c))$ and $1 - f(E(c))$ and store these values in λ_{Cat_1} and λ_{Tot}. These are used in reactions τ_3 and τ_4 to determine the probability that the steady state value of λ_{out} is going to be determined by reaction τ_5 or τ_6.

5.2   Compiling into the Class of NRO-CRSs

Given a formula P as defined in Definition 5, then $[\![P]\!]_E$ associates to P and an environment E a pmf. We now define a translation of P, $T(P)$, into the class of NRO-CRSs that guarantees that the unique output species of $T(P)$, at steady state, approximates $[\![P]\!]_E$ with arbitrarily small error for any environment E such that $V(P) \subseteq dom(E)$. In order to define such a translation we need the following renaming operator.

Definition 9. *Given a CRS $C = (\Lambda, R, x_0)$, for $\lambda_t \in \Lambda$ and $\lambda_1 \notin \Lambda$ we define the renaming operator $C\{\lambda_1 \leftarrow \lambda_t\} = C_c$ such that $C_c = ((\Lambda - \{\lambda_t\}) \cup \{\lambda_1\}, R\{\lambda_1 \leftarrow \lambda_t\}, x_0')$, where $R\{\lambda_1 \leftarrow \lambda_t\}$ substitutes any occurrence of λ_t with an occurrence of λ_1 for any $\tau \in R$ and $x_0'(\lambda) = \{x_0(\lambda) \text{ if } \lambda \neq \lambda_t; x_0(\lambda_t) \text{ if } \lambda = \lambda_1\}$.*

This operator produces a new CRS where any occurrence of a species is substituted with an occurrence of another species previously not present.

Definition 10 *(Translation into NRO-CRSs). Define the mapping T by induction on syntax of formulae P:*

$T(one) = (\{\lambda_{out}\}, \{\lambda_{out}\}, \emptyset, x_0)$ with $x_0(\lambda_{out}) = 1$;

$T(zero) = (\{\lambda_{out}\}, \{\lambda_{out}\}, \emptyset, x_0)$ with $x_0(\lambda_{out}) = 0$;

$T(P_1 + P_2) = Sum(T(P_1)\{\lambda_{o_1} \leftarrow \lambda_{out}\}, \lambda_{o_1}, T(P_2)\{\lambda_{o_2} \leftarrow \lambda_{out}\}, \lambda_{o_2}, \lambda_{out})$;

$T(k \cdot P) = Div(Mul(T(P)\{\lambda_o \leftarrow \lambda_{out}\}, \lambda_o, k_1, \lambda_{out})\{\lambda_{o'} \leftarrow \lambda_{out}\}), \lambda_{o'}, k_2, \lambda_{out})$;

$T(min(P_1, P_2) = Min(T(P_1)\{\lambda_{o_1} \leftarrow \lambda_{out}\}, \lambda_{o_1}, T(P_2)\{\lambda_{o_2} \leftarrow \lambda_{out}\}, \lambda_{o_2}, \lambda_{out})$;

$T((P_1)_D : P_2) =$

$$\begin{cases} Con(T(P_1)\{\lambda_{o_1} \leftarrow \lambda_{out}\}, \lambda_{o_1}, T(P_2)\{\lambda_{o_2} \leftarrow \lambda_{out}\}, \lambda_{o_2}, D, \lambda_{out}), & \text{if } D = p \\ ConE(T(P_1)\{\lambda_{o_1} \leftarrow \lambda_{out}\}, \lambda_{o_1}, T(P_2)\{\lambda_{o_2} \leftarrow \lambda_{out}\}, \lambda_{o_2}, D, \lambda_{out}), & \\ & \text{if } D = p + \sum_{i=1}^m p_i \cdot c_i \end{cases}$$

for $m > 1$, $k \in \mathbb{Q}_{>0}$, $k_1, k_2 \in \mathbb{N}$ such that $k = \frac{k_1}{k_2}$ and formulae P_1, P_2, which are assumed to not contain species $\lambda_{o_1}, \lambda_{o_2}$.

Example 5. Consider the formula $P_1 = (one)_{0.001 \cdot c + 0.2}(4 \cdot one) + (2 \cdot one)_{0.4}(3 \cdot one)$ of Example 2, and an environment E such that $0.000125 \leq E(c) \leq 1$ and suppose $E(c) \cdot 800 \in \mathbb{N}$. We show how the translation defined in Definition 10 produces a NRO-CRS C with output species λ_{out} such that $\pi_{\lambda_{out}} = [\![P_1]\!]_E$. Consider the following NRO-CRSs C_1, C_2, C_3, C_4 defined as $C_1 = (\{\lambda_{c_1}\}, \{\lambda_{c_1}\}, \{\}, x_0')$ with $x_0(\lambda_{c_1}) = 1$, $C_2 = (\{\lambda_{c_2}\}, \{\lambda_{c_2}\}, \{\}, x_0)$ with $x_0(\lambda_{c_2}) = 1,$, $C_3 = (\{\lambda_{c_3}\}, \{\lambda_{c_3}\}, \{\}, x_0)$ with $x_0(\lambda_{c_3}) = 1$, and $C_4 = (\{\lambda_{c_4}\}, \{\lambda_{c_4}\}, \{\}, x_0)$ with $x_0(\lambda_{c_2}) = 1$. Then, we have that:

$C_1^c = ConE(C_1, \lambda_{c_1}, Mul(C_2, \lambda_{c_2}, 4, \lambda_{out})\{\lambda_{o_2} \leftarrow \lambda_{out}\}, \lambda_{o_2}, 0.001 \cdot c + 0.2, \lambda_{out_1})$

$C_2^c = Con(Mul(C_3, \lambda_{c_3}, 2, \lambda_{out})\{\lambda_{o_3} \leftarrow \lambda_{out}\}, \lambda_{o_3},$

$\qquad\qquad\qquad Mul(C_4, \lambda_{c_4}, 3, \lambda_{out})\{\lambda_{o_4} \leftarrow \lambda_{out}\}, \lambda_{o_4}, 0.4, \lambda_{out_2})$

are such that $\pi_{\lambda_{out_1}} = \begin{cases} (0.001 \cdot [\![c]\!]_E + 0.2), & \text{if } y = 1 \\ 1 - (0.001 \cdot [\![c]\!]_E + 0.2), & \text{if } y = 4 \ , \\ 0, & \text{otherwise} \end{cases}$

and $\pi_{\lambda_{out_2}} = \begin{cases} 0.4, & \text{if } y = 2 \\ 0.6, & \text{if } y = 3 \ . \\ 0, & \text{otherwise} \end{cases}$ Then, consider the CRS $C = Sum(C_1^c$

$\{\lambda_{t_1} \leftarrow \lambda_{out_1}\}, \lambda_{t_1}, C_2^c\{\lambda_{t_2} \leftarrow \lambda_{out_2}\}, \lambda_{t_2}, \lambda_{out})$ and we have $\pi_{\lambda_{out}} = [\![P_1]\!]_E$ with arbitrarily small error. The reactions of C are shown below

Mul on inputs $\{\tau_1 : \lambda_{C_2} \rightarrow 4\lambda_{o_1}; \quad \tau_2 : \lambda_{C_3} \rightarrow 2\lambda_{o_2}; \quad \tau_3 : \lambda_{C_4} \rightarrow 3\lambda_{o_3}$

$C_1^c \begin{cases} \tau_4 : \lambda_{env} \rightarrow^k \lambda_{cat_1} + \lambda_{cat_2}; & \tau_5 : \lambda_{cat_1} + \lambda_z \rightarrow \lambda_1 \\ \tau_6 : \lambda_{cat_2} + \lambda_{tot} \rightarrow^k \emptyset; & \tau_7 : \lambda_{tot} + \lambda_z \rightarrow \lambda_2 \\ \tau_8 : \lambda_1 + \lambda_{o_1} \rightarrow \lambda_{o_1} + \lambda_{out_1}; & \tau_9 : \lambda_2 + \lambda_{o_2} \rightarrow \lambda_{o_2} + \lambda_{out_1} \end{cases}$

$C_2^c \begin{cases} \tau_9 : \lambda_{z_1} \rightarrow^{0.6} \lambda_{r_1}; & \tau_{10} : \lambda_{z_1} \rightarrow^{0.4} \lambda_{r_2} \\ \tau_{11} : \lambda_{r_1} + \lambda_{o_3}; \rightarrow \lambda_{r_1} + \lambda_{out_2}; & \tau_7 : \lambda_{r_2} + \lambda_{o_4} \rightarrow \lambda_{r_2} + \lambda_{out_2} \end{cases}$

$Sum \{\tau_{12} : \lambda_{out_1} \rightarrow \lambda_{out}; \quad \tau_{13} : \lambda_{out_2} \rightarrow \lambda_{out}$

for $k \gg 1$ and initial condition such that $x_0(\lambda_{env}) = E(c) \cdot 800$, $x_0(\lambda_{tot}) = 800$, $x_0(\lambda_z) = x_0(\lambda_{z_1}) = x_0(\lambda_{z_2}) = 1 = x_0(\lambda_{c_1}) = x_0(\lambda_{c_2}) = x_0(\lambda_{c_3}) = x_0(\lambda_{c_4}) = 1$, and all other species initialized with 0 molecules.

Proposition 2. *For any formula P we have that $T(P)$ is a NRO-CRS.*

The proof follows by structural induction as shown in [8]. Given a formula P and an environment E such that $V(P) \subseteq dom(E)$, the following theorem guarantees the soundness of $T(P)$ with respect to $[\![P]\!]_E$. In order to prove the soundness of our translation we consider the measure of the multiplicative error between two pmfs f_1 and f_2 with values in $\mathbb{N}^m$, $m > 0$ as $e_m(f_1, f_2) = \max_{n \in \mathbb{N}^m} \min(\frac{f_1(n)}{f_2(n)}, \frac{f_2(n)}{f_1(n)})$.

Theorem 5 *(Soundness). Given a formula P and λ_{out}, unique output species of $T(P)$, then, for an environment E such that $V(P) \subseteq dom(E)$, it holds that $\pi_{\lambda_{out}}^{T(P)} = [\![P]\!]_E$ with arbitrarily small error under multiplicative error measure.*

The proof follows by structural induction.

Remark 3. A formula P is finite by definition, so Theorem 5 is valid because the only production rule which can introduce an error is $(P_1)_D : (P_2)$ in the case $D \neq p_0$, and we can always find reaction rates to make the total probability of error arbitrarily small. Note that, by using the results of [17], it would also be possible to show that the total error can be kept arbitrarily small, even if a formula is composed from an unbounded number of production rules. This requires small modifications to the ConE operator following ideas in [17].

Note that compositional translation, as defined in Definition 10, generally produces more compact CRNs respect to the direct translation in Theorem 1, and in both cases the output is non-reacting, so the resulting CRN can be used for composition. For a distribution with support J direct translation yields a CRN with $2|J|$ reactions, whereas, for instance, the support of the sum pmf has the cardinality of the Cartesian product of the supports of the input pmfs.

6 Discussion

Our goal was to explore the capacity of CRNs to compute with distributions. This is an important goal because, when molecular interactions are in low number, as is common in various experimental scenarios [15], deterministic methods are not accurate, and stochasticity is essential for cellular circuits. Moreover, there is a large body of literature in biology where stochasticity has been shown to be essential and not only a nuisance [11]. Our work is a step forward towards better understanding of molecular computation. In this paper we focused on error-free computation for distributions. It would be interesting to understand and characterize what would happen when relaxing this constraint. That is, if we admit a probabilistically (arbitrarily) small error, does the ability of CRNs to compute on distributions increase? Can we relax the constraint that output species need to be produced monotonically? Can we produce more compact networks? Another topic we would like to address is if it is possible to implement the CRNs without leaders (species being present with initial number of molecules equal to 1). This is a crucial aspect in our theorems and having the same results without these constraints would make the implementation easier. However, it is worth noting that, in a practical scenario, such species could be thought of as a single gene or as localized structures [15].

References

1. Anderson, D.F., Craciun, G., Kurtz, T.G.: Product-form stationary distributions for deficiency zero chemical reaction networks. Bull. Math. Biol. **72**(8), 1947–1970 (2010)
2. Anderson, D.F., Kurtz, T.G.: Continuous time Markov chain models for chemical reaction networks. In: Koeppl, H., Setti, G., di Bernardo, M., Densmore, D. (eds.) Design and Analysis of Biomolecular Circuits, pp. 3–42. Springer, New York (2011)
3. Anderson, J.C., Clarke, E.J., Arkin, A.P., Voigt, C.A.: Environmentally controlled invasion of cancer cells by engineered bacteria. J. Mol. Biol. **355**(4), 619–627 (2006)
4. Angluin, D., Aspnes, J., Eisenstat, D., Ruppert, E.: The computational power of population protocols. Distrib. Comput. **20**(4), 279–304 (2007)
5. Arkin, A., Ross, J., McAdams, H.H.: Stochastic kinetic analysis of developmental pathway bifurcation in phage λ-infected Escherichia coli cells. Genetics **149**(4), 1633–1648 (1998)
6. Cardelli, L., Csikász-Nagy, A.: The cell cycle switch computes approximate majority. Sci. Rep. **2** (2012)
7. Cardelli, L., Kwiatkowska, M., Laurenti, L.: Stochastic analysis of chemical reaction networks using linear noise approximation. In: Roux, O., Bourdon, J. (eds.) CMSB 2015. LNCS, vol. 9308, pp. 64–76. Springer, Heidelberg (2015)
8. Cardelli, L., Kwiatkowska, M., Laurenti, L.: Programming discrete distributions with chemical reaction networks (2016). arXiv preprint arXiv:1601.02578
9. Chen, H.-L., Doty, D., Soloveichik, D.: Deterministic function computation with chemical reaction networks. Natural Comput. **13**(4), 517–534 (2014)
10. Chen, Y.-J., Dalchau, N., Srinivas, N., Phillips, A., Cardelli, L., Soloveichik, D., Seelig, G.: Programmable chemical controllers made from DNA. Nat. Nanotechnol. **8**(10), 755–762 (2013)

11. Eldar, A., Elowitz, M.B.: Functional roles for noise in genetic circuits. Nature **467**(7312), 167–173 (2010)
12. Fett, B., Bruck, J., Riedel, M.D.: Synthesizing stochasticity in biochemical systems. In: 2007 44th ACM/IEEE Design Automation Conference, DAC 2007, pp. 640–645. IEEE (2007)
13. Kwiatkowska, M., Norman, G., Parker, D.: Stochastic model checking. In: Bernardo, M., Hillston, J. (eds.) SFM 2007. LNCS, vol. 4486, pp. 220–270. Springer, Heidelberg (2007)
14. Losick, R., Desplan, C.: Stochasticity and cell fate. Science **320**(5872), 65–68 (2008)
15. Qian, L., Winfree, E.: Parallel and scalable computation and spatial dynamics with DNA-based chemical reaction networks on a surface. In: Murata, S., Kobayashi, S. (eds.) DNA 2014. LNCS, vol. 8727, pp. 114–131. Springer, Heidelberg (2014)
16. Schmiedel, J.M., Klemm, S.L., Zheng, Y., Sahay, A., Blüthgen, N., Marks, D.S., van Oudenaarden, A.: MicroRNA control of protein expression noise. Science **348**(6230), 128–132 (2015)
17. Soloveichik, D., Cook, M., Winfree, E., Bruck, J.: Computation with finite stochastic chemical reaction networks. Nat. Comput. **7**(4), 615–633 (2008)
18. Soloveichik, D., Seelig, G., Winfree, E.: DNA as a universal substrate for chemical kinetics. Proc. Nat. Acad. Sci. **107**(12), 5393–5398 (2010)

Robustness of Expressivity in Chemical Reaction Networks

Robert Brijder[1]([✉]), David Doty[2], and David Soloveichik[3]

[1] Hasselt University, Diepenbeek, Belgium
robert.brijder@uhasselt.be
[2] University of California, Davis, CA, USA
doty@ucdavis.edu
[3] University of Texas, Austin, TX, USA
david.soloveichik@utexas.edu

Abstract. We show that some natural output conventions for error-free computation in chemical reaction networks (CRN) lead to a common level of computational expressivity. Our main results are that the standard definition of error-free CRNs have equivalent computational power to (1) *asymmetric* and (2) *democratic* CRNs. The former have only "yes" voters, with the interpretation that the CRN's output is yes if any voters are present and no otherwise. The latter define output by majority vote among "yes" and "no" voters.

Both results are proven via a generalized framework that simultaneously captures several definitions, directly inspired by a recent Petri net result of Esparza, Ganty, Leroux, and Majumder [CONCUR 2015]. These results support the thesis that the computational expressivity of error-free CRNs is intrinsic, not sensitive to arbitrary definitional choices.

1   Introduction

Turing machines solve exactly the same class of yes/no decision problems whether they report output via accept/reject states, or if instead they write a 1 or 0 on a worktape before halting. Similarly, finite-state transducers compute the same class of functions whether they emit output on a state (*Moore machine* [19]) or a transition (*Mealy machine* [18]). In general, if the power of a model of computation is insensitive to minor changes in the definition, this lends evidence to the claim that the model is robust enough to apply to many real situations, and that theorems proven in the model reflect fundamental truths about reality, rather than being artifacts of arbitrary definitional choices.

The theory of chemical reaction networks (CRNs) studies the general behavior of chemical reactions in well-mixed solutions, abstracting away spatial properties of the molecules. Formally, a CRN is defined as a finite set of reactions

The first author is a postdoctoral fellow of the Research Foundation – Flanders (FWO). The second author was supported by NSF grant 1619343, and the third author by NSF grant 1618895.

© Springer International Publishing Switzerland 2016
Y. Rondelez and D. Woods (Eds.): DNA 2016, LNCS 9818, pp. 52–66, 2016.
DOI: 10.1007/978-3-319-43994-5_4

such as $2A + C \rightarrow 2B$, where A, B, and C are abstract chemical species. In a discrete CRN the state of the system is given by molecule counts of each species and the system updates by application of individual reactions.

CRNs have only recently been considered as a model of computation [21], motivated partially by the ability to implement them using a basic experimental technique called *DNA strand displacement* [22]. Discrete CRNs are Turing complete if allowed an arbitrary small, but nonzero, probability of error [21], improved to probability 0 in [9]. Using a result from the theory of population protocols [3,4], it is known that error-free CRNs decide exactly the semilinear sets [6].[1]

We study the computational robustness of error-free CRNs under different output conventions. The original output convention [3] for deciding predicates (0/1-valued functions) is that each species is classified as voting either 0 ("no") or 1 ("yes"), and a configuration (vector of nonnegative integer counts of each species) o has output $i \in \{0,1\}$ if all species present in positive count are i-voters, i.e., there is a *consensus* on vote i. As an example, the CRN with reactions $X_1 + N \rightarrow Y$ and $X_2 + Y \rightarrow N$, with initial configuration $\{x_1 X_1, x_2 X_2, 1N\}$, where N, X_2 vote 0 and Y, X_1 vote 1, decides if $x_1 > x_2$; Y and N alternate being present as each reacts with an input, so the first input to run out determines whether we stop at Y or N. More formally, we say o is *output-stable* if every configuration o' reachable from o has the same output as o (i.e., the system need not halt, but it stops changing its output). Finally, it is required that a correct output-stable configuration is reachable not only from the initial configuration i, but also from any configuration reachable from i; under mild assumptions (e.g., conservation of mass), this implies that a correct stable configuration is actually reached with probability 1 under the standard stochastic kinetic model [14]. It has been shown in [3] that the computational power is not reduced, that is, it still decides precisely all semilinear sets, when we restrict to those CRNs where (1) each reaction has two reactants and two products (e.g., disallowing reactions such as $2A+C \rightarrow 2B$ and $A \rightarrow B+C$, a model known as a *population protocol* [3]) and (2) the system eventually halts for every possible input (see also [7]).

One can imagine alternative output conventions, i.e., ways to interpret what is the output of a configuration, while retaining the requirement that a correct output-stable configuration is reachable from any reachable configuration. Rather than requiring every species to vote 0 or 1, for example, allow the CRN to designate some species as nonvoters. It is not difficult to show that such CRNs have equivalent computational power: They are at least as powerful since one can always choose all species to be voters. The reverse direction follows by converting a CRN with a subset of voting species into one in which every species

---

[1] We use the term "error-free" in this section to refer to a specific requirement of "stability" defined formally in Sect. 2.2. When the set of configurations reachable from an initial configuration is always finite (for instance, with population protocols, or more generally mass-conserving CRNs), then stability coincides with probability 0 of error. See [9] for an in-depth discussion of how these notions can diverge when the set of configurations reachable from an initial configuration is infinite.

votes, by replacing every nonvoting species S with two variants S_0 and S_1, whose voting bit is swayed by reactions with the original voting species, and which are otherwise both functionally equivalent to S.

We investigate two output conventions that are not so easily seen to be convertible to the original convention. The first convention is the *asymmetric* model, in which there are only 1-voters, whose presence or absence indicates a configuration-wide output of 1 or 0, respectively. It is not obvious how to convert an asymmetric CRN into a symmetric CRN, since this appears to require producing 0-voters if and only if 1-voters are absent. The second convention is the *democratic* model, in which there are 0- and 1-voters, but the output of a configuration is given by the majority vote rather than being defined only with consensus. Intuitively, the difficulty in converting a democratic CRN into a symmetric consensus CRN is that, although the democratic CRN may stabilize on a majority of, for example, 1-voters over 0-voters, the exact numerical gap between them may never stabilize. A straightforward attempt to convert a democratic CRN into a consensus CRN results in a CRN that changes the output every time a new 0- or 1-voter appears. For instance, suppose we use the previously described CRN for computing whether $x_1 > x_0$, where x_1 and x_0 respectively represent the count of 1- and 0-voters. If the original democratic CRN repeatedly increments x_0 and then x_1, the resulting CRN flips between Y and N indefinitely — thus never stabilizing in the consensus model — even if $x_1 > x_0$ remains true indefinitely.

We show that these conventions have equivalent power as the original definition. Our techniques further establish that the class of predicates computable by CRNs is robust to two additional relaxations of the classical notion of stable computation [3]: (1) a correct output configuration need not be reachable from *every* reachable configuration, only the initial configuration, and (2) the set of output configurations need not be "stable" (i.e., closed under application of reactions), so long as each initial configuration can reach only a correct output.

After defining existing notions of computation by CRNs in Sect. 2, we introduce in Sect. 3 a very general computational model for CRNs, called a *generalized chemical reaction decider* (gen-CRD). Its definition is directly inspired by a recent powerful result from Petri net theory [13], restated here as Theorem 3.2. Using this result we show that under mild conditions, gen-CRDs decide only semilinear sets. We then show that the original symmetric consensus model, the asymmetric consensus model, and the symmetric majority model all fit into this framework, establishing their common expressivity.

2   Chemical Reaction Networks and Deciders

2.1   Chemical Reaction Networks

Let $\mathbb{Z}$ and $\mathbb{N}$ denote the integers and nonnegative integers, respectively. Let Λ be a finite set. The set of vectors over $\mathbb{N}$ indexed by Λ (i.e., the set of functions $c : \Lambda \to \mathbb{N}$) is denoted by $\mathbb{N}^\Lambda$. The zero vector is denoted $\mathbf{0}$. For $c, c' \in \mathbb{N}^\Lambda$ we write $c \leq c'$ if and only if $c(S) \leq c'(S)$ for all $S \in \Lambda$. For $c \in \mathbb{N}^\Lambda$ and

$\Sigma \subseteq \Lambda$, the *projection* of c to Σ, denoted by $c \upharpoonright_\Sigma$, is an element in $\mathbb{N}^\Sigma$ such that $c \upharpoonright_\Sigma (S) = c(S)$ for all $S \in \Sigma$. Let $\|c\| = \|c\|_1 = \sum_{S \in \Lambda} c(S)$ denote the L_1 norm of c. We sometimes use multiset notation, e.g., $c = \{1A, 2C\}$ to denote $c(A) = 1, c(C) = 2, c(S) = 0$ for $S \notin \{A, C\}$, or when defining reactions, additive notation, i.e., $A + 2C$.

A *reaction* α over Λ is a pair (r, p) with $r, p \in \mathbb{N}^\Lambda$ and $r \neq p$, where r and p are the *reactants* and *products* of α, respectively. We write $r \to p$ to denote a reaction (r, p), e.g., $A + B \to 2A + C$ denotes the reaction $(\{A, B\}, \{2A, C\})$.

Definition 2.1. *A chemical reaction network (CRN) is an ordered pair $\mathcal{N} = (\Lambda, R)$ with Λ a finite set and R a finite set of reactions over Λ.*

The elements of Λ are called the *species* of $\mathcal{N}$. The elements of $\mathbb{N}^\Lambda$ are called the *configurations* of $\mathcal{N}$. Viewing c as a multiset, each element of c is called a *molecule*. For $c, c' \in \mathbb{N}^\Lambda$, we write $c \Rightarrow_\mathcal{N} c'$ if there is a reaction $\alpha = (r, p) \in R$ such that $r \leq c$ and $c' = c - r + p$. The transitive and reflexive closure of $\Rightarrow_\mathcal{N}$ is denoted by $\Rightarrow_\mathcal{N}^*$. If $\mathcal{N}$ is clear from the context, then we simply write $\Rightarrow$ and $\Rightarrow^*$ for $\Rightarrow_\mathcal{N}$ and $\Rightarrow_\mathcal{N}^*$, respectively. If $c \Rightarrow^* c'$, then we say c' is *reachable* from c.

For $c \in \mathbb{N}^\Lambda$, we define $\mathsf{pre}_\mathcal{N}(c) = \{c' \in \mathbb{N}^\Lambda \mid c' \Rightarrow_\mathcal{N}^* c\}$ and $\mathsf{post}_\mathcal{N}(c) = \{c' \in \mathbb{N}^\Lambda \mid c \Rightarrow_\mathcal{N}^* c'\}$. Again we omit the subscript $\mathcal{N}$ if the CRN $\mathcal{N}$ is clear from the context. Note that for $c, c' \in \mathbb{N}^\Lambda$, we have $c \in \mathsf{pre}(c')$ if and only if $c' \in \mathsf{post}(c)$ if and only if $c \Rightarrow^* c'$. We extend $\mathsf{pre}(c)$ and $\mathsf{post}(c)$ to sets $X \subseteq \mathbb{N}^\Lambda$ in the natural way: $\mathsf{pre}(X) = \bigcup_{c \in X} \mathsf{pre}(c)$ and $\mathsf{post}(X) = \bigcup_{c \in X} \mathsf{post}(c)$.

Petri net theory is a very well established theory of concurrent computation [20]. We recall here that CRNs are essentially equivalent to Petri nets. In Petri net terminology, molecules are called "tokens", species are called "places", reactions are called "transitions", and configurations are called "markings". Due to this correspondence, we can apply results from Petri net theory to CRNs (which we will do in this paper, cf. Theorem 3.2). Conversely, the results shown in this paper can be reformulated straightforwardly in terms of Petri nets. Vector addition systems [17] form a model nearly equivalent to CRNs and Petri nets, where reactions roughly correspond to vectors with integer entries.[2] In the special case of population protocols [3], each reaction $\alpha = (r, p)$ obeys $\|r\| = \|p\| = 2$. As a result, for each configuration c of a population protocol, both $\mathsf{pre}(c)$ and $\mathsf{post}(c)$ are finite (because there are only a finite number of configurations c' with $\|c'\| = \|c\|$). In that model, molecules are called "agents", species are called "states", and reactions are called "transitions".

2.2   Symmetric Output-Stable Deciders

We now recall how one can compute using CRNs. Say we want to decide whether or not the number n of molecules of species X is even. One way to do this is by

---

[2] The only difference is *catalysts*: reactants that are also products, e.g., $C + X \to C + Y$, are allowed in CRNs and Petri nets but not in vector addition systems. Most results for these models are insensitive to this difference.

introducing the reaction $X + X \to \varnothing$.[3] If n is even, then eventually all molecules are consumed, and if n is odd, then eventually there is exactly one molecule of species X present. Once the CRN has stabilized, the presence of a molecule of species X signals that n is odd (i.e., there were an odd number of molecules of species X present initially). Note that in this example there is no molecule of any species that signals that n is even. One may think of a more elaborate example where the presence of say, a molecule of species V_{even}, signals (once the CRN has stabilized) that n is even. In this way, once the CRN has stabilized, X "votes" that n is odd, while V_{even} "votes" that n is even.

A chemical reaction decider $\mathcal{D}$ (introduced in [8]) is a reformulation in terms of CRNs of the notion of population protocol [3] from the field of distributed computing. We define a set of input configurations $\mathcal{I}$ and two sets of "trap configurations", called *output-stable* configurations, $\mathcal{O}_0$ and $\mathcal{O}_1$. We then say that $\mathcal{D}$ is *output-stable* and *decides* the set $\mathcal{I}_1 \subseteq \mathcal{I}$ (with $\mathcal{I}_0 = \mathcal{I} \setminus \mathcal{I}_1$) if for each $i \in \{0,1\}$ (1) starting from a configuration in $\mathcal{I}_i$, the CRN remains always within reach of a configuration in $\mathcal{O}_i$ (i.e., $\mathsf{post}(\mathcal{I}_i) \subseteq \mathsf{pre}(\mathcal{O}_i)$), and (2) once a configuration is in $\mathcal{O}_i$, it is stuck in $\mathcal{O}_i$ (i.e., $\mathsf{post}(\mathcal{O}_i) = \mathcal{O}_i$).

The sets $\mathcal{I}$, $\mathcal{O}_0$, and $\mathcal{O}_1$ are all of a specific form. There is a subset of *input species* $\Sigma \subseteq \Lambda$; $\mathcal{I}$ consists of nonzero configurations where the all molecules present are in Σ. The output is based on consensus: all the molecules present in an output configuration must agree on the output. More precisely, there is a partition $\{\Gamma_0, \Gamma_1\}$ of Λ (called *0-voters* and *1-voters*, respectively),[4] such that configuration $\boldsymbol{c}$ has *output* $i \in \{0,1\}$ if all molecules present in $\boldsymbol{c}$ are from Γ_i (i.e., $\boldsymbol{c}{\restriction}_{\Gamma_{1-i}} = \boldsymbol{0}$) and $\boldsymbol{c} \neq \boldsymbol{0}$). A configuration $\boldsymbol{o}$ is defined to be in $\mathcal{O}_i$ — it is *output-stable* — if all configurations of $\mathsf{post}(\boldsymbol{o})$ also have output i.

Our definition, though equivalent, is phrased differently from the usual one [3], being defined in terms of $\mathcal{I}$, $\mathcal{O}_0$, and $\mathcal{O}_1$ instead of Σ, Γ_0, and Γ_1. This simplifies our generalization of this notion in Sect. 3.

Definition 2.2. *A symmetric output stable chemical reaction decider (sym-CRD) is a 4-tuple* $\mathcal{D} = (\mathcal{N}, \mathcal{I}, \mathcal{O}_0, \mathcal{O}_1)$, *where* $\mathcal{N} = (\Lambda, R)$ *is a CRN and there are* $\Sigma \subseteq \Lambda$ *and a partition* $\{\Gamma_0, \Gamma_1\}$ *of* Λ *such that*

1. $\mathcal{I} = \{\boldsymbol{c} \in \mathbb{N}^{\Lambda} \mid \boldsymbol{c}{\restriction}_{\Lambda \setminus \Sigma} = \boldsymbol{0}\} \setminus \{\boldsymbol{0}\}$,
2. $\mathcal{O}_i = \{\boldsymbol{c} \in \mathbb{N}^{\Lambda} \mid \mathsf{post}(\boldsymbol{c}) \subseteq \mathcal{L}_i \setminus \mathcal{L}_{1-i}\}$, *with* $\mathcal{L}_i = \{\boldsymbol{c} \in \mathbb{N}^{\Lambda} \mid \boldsymbol{c}{\restriction}_{\Gamma_i} \neq \boldsymbol{0}\}$ *for* $i \in \{0,1\}$.
3. *There is a partition* $\{\mathcal{I}_0, \mathcal{I}_1\}$ *of* $\mathcal{I}$ *such that* $\mathsf{post}(\mathcal{I}_i) \subseteq \mathsf{pre}(\mathcal{O}_i)$ *for* $i \in \{0,1\}$.

Condition 1 states that only species in Σ may be present initially, and at least one must be present. Condition 2 defines $\mathcal{L}_i$ to be configurations with an i-voter, so those in $\mathcal{L}_i \setminus \mathcal{L}_{1-i}$ unanimously vote i, and those in $\mathcal{O}_i$ are stable ("stuck" in the set $\mathcal{L}_i \setminus \mathcal{L}_{1-i}$). Condition 3 states that from every configuration

---

[3] Notation $\varnothing$ indicates that this reaction has no products.

[4] The definition of [8] allows only a subset of Λ to be voters, i.e., $\Gamma_0 \cup \Gamma_1 \subseteq \Lambda$. This convention is more easily shown to define equivalent computational power than our main results about asymmetric and democratic voting.

reachable from an initial configuration, a "correct" output stable configuration is reachable from there; this is the usual way of expressing stable computation [6,8]. The relationships between these sets are depicted in Fig. 1.

Remark 2.3. A different definition is found in [8] and a number of other papers. That definition relaxes ours in two ways: (1) having both voting and non-voting species, (2) allowing non-input species in the input configuration (e.g., $\{1N\}$ in the Introduction). It turns out that (1) does not affect the computational power of the model. It is also known [3] that (2) does not alter the computational power (though it may affect the time complexity [5,12]).

Remark 2.4. We can equivalently define $\mathcal{O}_i = \mathbb{N}^\Lambda \setminus \mathsf{pre}(\mathcal{L}_{1-i} \cup \{\mathbf{0}\})$, a form that will be useful later. To see that this definition is equivalent, observe that $\mathbb{N}^\Lambda \setminus \mathcal{O}_i$ is the set of configurations from which it is possible either to reach $\mathcal{L}_{1-i}$, or to reach *outside* of $\mathcal{L}_i$, and the only point outside *both* is $\mathbf{0}$, so $\mathbb{N}^\Lambda \setminus \mathcal{O}_i = \mathsf{pre}(\mathcal{L}_{1-i} \cup \{\mathbf{0}\})$. Thus $\mathcal{O}_i = \mathbb{N}^\Lambda \setminus \mathsf{pre}(\mathcal{L}_{1-i} \cup \{\mathbf{0}\})$.

Remark 2.5. The $\mathcal{O}_i$ are disjoint and closed under application of reactions: $\mathcal{O}_0 \cap \mathcal{O}_1 = \varnothing$ and $\mathsf{post}(\mathcal{O}_i) = \mathcal{O}_i$.

Remark 2.6. Definition 2.2 implies the (weaker) condition that $\mathcal{I}_i = \mathcal{I} \cap \mathsf{pre}(\mathcal{O}_i)$. This can be shown as follows. First, $\mathcal{I}_i \subseteq \mathcal{I}$ and $\mathcal{I}_i \subseteq \mathsf{post}(\mathcal{I}_i) \subseteq \mathsf{pre}(\mathcal{O}_i)$, so $\mathcal{I}_i \subseteq \mathcal{I} \cap \mathsf{pre}(\mathcal{O}_i)$. To see the reverse containment, assume $c \in \mathcal{I} \cap \mathsf{pre}(\mathcal{O}_i)$, but $c \notin \mathcal{I}_i$, i.e., $c \in \mathcal{I}_{1-i} \cap \mathsf{pre}(\mathcal{O}_i)$. Let $o \in \mathsf{post}(c)$ be such that $o \in \mathcal{O}_i$; such o exists since $c \in \mathsf{pre}(\mathcal{O}_i)$. Since $o \in \mathsf{post}(\mathcal{I}_{1-i}) \subseteq \mathsf{pre}(\mathcal{O}_{1-i})$, we have $o \in \mathcal{O}_i \cap \mathsf{pre}(\mathcal{O}_{1-i})$. Let $o' \in \mathsf{post}(o)$ such that $o' \in \mathcal{O}_{1-i}$. Then $o' \in \mathsf{post}(\mathcal{O}_i) \cap \mathcal{O}_{1-i}$ — a contradiction because $\mathsf{post}(\mathcal{O}_i) = \mathcal{O}_i$ is disjoint from $\mathcal{O}_{1-i}$.

Since $\mathcal{I}_0 = \mathcal{I} \cap \mathsf{pre}(\mathcal{O}_0)$ and $\mathcal{I}_1 = \mathcal{I} \cap \mathsf{pre}(\mathcal{O}_1)$ are disjoint, we say that a sym-CRD $\mathcal{D}$ *decides* the set $\mathcal{I}_1$. If a sym-CRD $\mathcal{D}$ decides the set $X \subseteq \mathbb{N}^\Lambda$, then the entries indexed by $\Lambda \setminus \Sigma$ are zero for each $c \in X$. Therefore, by abuse of notation, we also say that $\mathcal{D}$ *decides* the set $X{\restriction_\Sigma} \subseteq \mathbb{N}^\Sigma$. We will use this convention for all chemical reaction deciders with $\mathcal{I}$ of the given form.

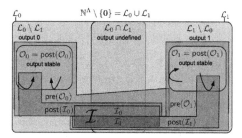

Fig. 1. Venn diagram of configurations that define sym-CRD. Subset relationships depicted in their most general form: $\mathcal{I}_i \subseteq \mathsf{post}(\mathcal{I}_i) \subseteq \mathsf{pre}(\mathcal{O}_i)$, and $\mathcal{O}_i \subseteq \mathcal{L}_i \setminus \mathcal{L}_{1-i}$. $\mathsf{pre}(\mathcal{O}_0)$ and $\mathsf{pre}(\mathcal{O}_1)$ partition the set $\mathcal{I} = \mathcal{I}_0 \cup \mathcal{I}_1$.

Example 2.7. We construct a sym-CRD $\mathcal{D}$ that decides the set $x \not\equiv y$ mod m where x and y are non-negative integer variables, not both zero, and $m \geq 2$ is an integer constant. The variables x and y represent initial counts of species X and Y, respectively. Let $\Sigma = \{X, Y\}$, $\Gamma_0 = \{V_0\}$, $\Gamma_1 = \{X, Y\}$, and $\Lambda = \Gamma_0 \cup \Gamma_1$ be as in Definition 2.2, with the following reactions:

$$mX \to V_0, \quad mY \to V_0, \quad X + Y \to V_0, \tag{2.1}$$

$$Y + V_0 \to Y, \quad X + V_0 \to X. \tag{2.2}$$

We argue that $\mathcal{D}$ decides the set $\{c \in \mathbb{N}^\Sigma \setminus \{0\} \mid c(X) \not\equiv c(Y) \bmod m\}$. Indeed, if $x \equiv y \bmod m$, then eventually all X and Y molecules are consumed by the reactions of (2.1). The last time one of these reactions occurs introduces a V_0 molecule (there is a last reaction since x and y are not both zero). So eventually we obtain a configuration $c \in \mathcal{L}_0 \setminus \mathcal{L}_1$ for which no reaction can be applied anymore. Thus $c \in \mathcal{O}_0$. If $x \not\equiv y \bmod m$, then eventually we reach a configuration with one of X or Y, but not both, remaining. The remaining X or Y molecules consume all V_0 molecules by the reactions of (2.2), without the possibility of producing any more. So eventually we obtain a configuration $c' \in \mathcal{L}_1 \setminus \mathcal{L}_0$ for which no reaction can be applied anymore. Thus $c' \in \mathcal{O}_1$.

2.3 Semilinear Sets

We say that $X \subseteq \mathbb{N}^\Lambda$ is *linear* if there is a finite set $\{v_1, \ldots, v_k\} \subseteq \mathbb{N}^\Lambda$ and $b \in \mathbb{N}^\Lambda$ such that $X = \{b + \sum_{i=1}^k n_i v_i \mid n_1, \ldots, n_k \in \mathbb{N}\}$. We say that $X \subseteq \mathbb{N}^\Lambda$ is *semilinear* if X is the union of a finite number of linear sets. Semilinear sets are precisely the sets definable in Presburger arithmetic, which is the first-order theory of natural numbers with addition. As a consequence, the class of semilinear sets is closed under union, intersection, complementation, and projection [15]. An useful characterization of semilinear sets is that they are exactly the sets expressible as finite unions, intersections, and complements of sets of one of the following two forms: *threshold sets* of the form $\{x \mid \sum_{i=1}^k a_i \cdot x(i) < b\}$ for some constants $a_1, \ldots, a_k, b \in \mathbb{Z}$ or *mod sets* of the form $\{x \mid \sum_{i=1}^k a_i \cdot x(i) \equiv b \bmod c\}$ for some constants $a_1, \ldots, a_k \in \mathbb{Z}$ and $b, c \in \mathbb{N}$.

The following result was shown in [3,4]. In fact, the result was shown for output-stable population protocols, which form a subclass of the sym-CRDs. However, the proof is sufficiently general to hold for sym-CRDs as well.[5]

Theorem 2.8 [3,4]. *Let $X \subseteq \mathbb{N}^\Sigma \setminus \{0\}$. Then X is semilinear if and only if there is a sym-CRD that decides X.*

For a configuration $c \in \mathbb{N}^\Sigma$, $\mathsf{pre}(c)$ and $\mathsf{post}(c)$ are in general *not* semilinear [16]. Hence the semilinearity of Theorem 2.8 is due to additional "computational structure" of a sym-CRD. We repeatedly use the following notion of upwards closure to prove that certain sets are semilinear. The results below were shown or implicit in earlier papers [4,10]. We say $X \subseteq \mathbb{N}^\Lambda$ is *closed upwards* if, for all $c \in X$, $c' \geq c$ implies $c' \in X$.

Lemma 2.9. *Every closed upwards set $X \subseteq \mathbb{N}^\Lambda$ is semilinear.*

---

[5] Indeed, the negative result of [4] that sym-CRDs decide only semilinear sets is more general than stated in Theorem 2.8, applying to *any* reachability relation $\Rightarrow^*$ on $\mathbb{N}^\Lambda$ that is reflexive, transitive, and "additive" ($x \Rightarrow^* y$ implies $x + c \Rightarrow^* y + c$). Also, the negative result of [4] implicitly assumes that the zero vector 0 is not reachable (i.e., $\mathsf{pre}(0) = \{0\}$). This assumption is manifest for population protocols (if the population size is non-zero). For CRNs, this assumption can be readily removed; see Lemma 2.11.

Lemma 2.10. *If $X \subseteq \mathbb{N}^\Lambda$ is closed upwards, then so are $\mathsf{pre}(X)$ and $\mathsf{post}(X)$.*

Our results require $\mathsf{pre}(\mathbf{0})$ to be semilinear. Observe that $\mathsf{pre}(\mathbf{0}) = \{\mathbf{0}\}$ if and only if for each reaction $\alpha = (\boldsymbol{r}, \boldsymbol{p})$, $\boldsymbol{p} \neq \mathbf{0}$. The next lemma shows that we can assume this holds for sym-CRDs without loss of generality.

Lemma 2.11. *For every sym-CRD $\mathcal{D}$, there is a sym-CRD $\mathcal{D}'$ deciding the same set such that, for each reaction $\alpha = (\boldsymbol{r}, \boldsymbol{p})$ of $\mathcal{D}'$, $\boldsymbol{p} \neq \mathbf{0}$.*

3   Generalized Chemical Reaction Deciders

In this section, we formulate a more generalized definition of CRDs that captures the original "symmetric" definition (sym-CRD) in Sect. 2.2 and the new "asymmetric" definition (asym-CRD) in Sect. 4, as well as the "democratic" definition (dem-CRD) in Sect. 5. In this section we show how to use a result of [13] to re-prove the result of Angluin, Aspnes, and Eisenstat [4] that sym-CRDs decide only semilinear sets. This is a warmup to our main results, shown in Sects. 4 and 5, that asym-CRDs and dem-CRDs decide exactly the semilinear sets.

In the generalized notion defined below we have dropped the specific structure of $\mathcal{I}$, $\mathcal{O}_0$, and $\mathcal{O}_1$ (they are now arbitrary subsets of $\mathbb{N}^\Lambda$) and we have replaced the requirement that $\mathsf{post}(\mathcal{I}_i) \subseteq \mathsf{pre}(\mathcal{O}_i)$ by the weaker condition that $\mathcal{I}_i = \mathcal{I} \cap \mathsf{pre}(\mathcal{O}_i)$ (recall Remark 2.6). Also, we do not use the term "stable" in reference to this generalized notion, since there is no requirement that the output configurations $\mathcal{O}_i$ be closed under application of reactions (i.e., we allow $\mathcal{O}_i \subsetneq \mathsf{post}(\mathcal{O}_i)$).

The relationships among the sets relevant to the definition below are depicted in Fig. 2.

Definition 3.1. *A generalized chemical reaction decider (gen-CRD) is a*

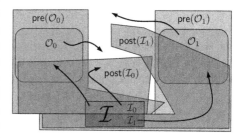

Fig. 2. Venn diagram of configurations that define generalized chemical reaction decider (gen-CRD). Like sym-CRD, $\mathsf{pre}(\mathcal{O}_0)$ and $\mathsf{pre}(\mathcal{O}_1)$ partition the input set $\mathcal{I} = \mathcal{I}_0 \cup \mathcal{I}_1$. Differences with sym-CRD: 1) Possibly $\mathcal{O}_i \subsetneq \mathsf{post}(\mathcal{O}_i)$ (output is not necessarily "stable"). 2) Although $\mathcal{I}_i \subseteq \mathsf{pre}(\mathcal{O}_i)$ (correct output reachable initially), yet possibly $\mathsf{post}(\mathcal{I}_i) \not\subseteq \mathsf{pre}(\mathcal{O}_i)$ (correct output could become unreachable).

4-tuple $\mathcal{D} = (\mathcal{N}, \mathcal{I}, \mathcal{O}_0, \mathcal{O}_1)$, where $\mathcal{N} = (\Lambda, R)$ is a CRN, $\mathcal{I}, \mathcal{O}_0, \mathcal{O}_1 \subseteq \mathbb{N}^\Lambda$, and there is a partition $\{\mathcal{I}_0, \mathcal{I}_1\}$ of $\mathcal{I}$ such that $\mathcal{I}_i = \mathcal{I} \cap \mathsf{pre}(\mathcal{O}_i)$ for $i \in \{0, 1\}$.

Observe that every sym-CRD is a gen-CRD. However, the requirements to be a gen-CRD are weaker than for sym-CRDs: (1) the condition $\mathsf{post}(\mathcal{O}_i) = \mathcal{O}_i$ need not hold for gen-CRDs, so it may be possible to "escape" from $\mathcal{O}_i$, and

(2) since $\mathsf{post}(\mathcal{I}_i) \subseteq \mathsf{pre}(\mathcal{O}_i)$ need not hold for gen-CRDs, it is possible to take a "wrong" route starting from $\mathcal{I}_i$ such that $\mathcal{O}_i$ becomes unreachable.[6]

Despite these relaxations, observe that the following property of sym-CRDs is retained in gen-CRDs: $\mathcal{I}$ is the disjoint union of $\mathcal{I}_0 = \mathcal{I} \cap \mathsf{pre}(\mathcal{O}_0)$ and $\mathcal{I}_1 = \mathcal{I} \cap \mathsf{pre}(\mathcal{O}_1)$, *i.e.*, from each input configuration, *exactly one* of the two output sets $\mathcal{O}_0$ or $\mathcal{O}_1$ is reachable. We say that a gen-CRD $\mathcal{D}$ *decides* the set $\mathcal{I}_1$.

Definition 3.1 is inspired by the following key Petri net result from [13, Theorem 10] (formulated here in terms of CRNs).

Theorem 3.2 [13]. *Let $\mathcal{N}$ be a CRN and $\mathcal{O}_0, \mathcal{O}_1, \mathcal{I} \subseteq \mathbb{N}^\Lambda$ be semilinear. If $\{\mathcal{I}_0, \mathcal{I}_1\}$ is a partition of $\mathcal{I}$ with $\mathcal{I}_i = \mathcal{I} \cap \mathsf{pre}(\mathcal{O}_i)$ for $i \in \{0,1\}$, then $\mathcal{I}_0$ and $\mathcal{I}_1$ are semilinear.*

We say that a gen-CRD $\mathcal{D} = (\mathcal{N}, \mathcal{I}, \mathcal{O}_0, \mathcal{O}_1)$ is *semilinear* if $\mathcal{I}$, $\mathcal{O}_0$, and $\mathcal{O}_1$ are all semilinear. We immediately have the following corollary to Theorem 3.2.

Corollary 3.3. *If a semilinear gen-CRD decides $X \subseteq \mathbb{N}^\Lambda$, then X is semilinear.*

As a by-product of the results shown in [13], the reverse direction of Theorem 2.8 (which is the most difficult implication) was reproven in [13] for the case of population protocols. That proof however essentially uses the fact that, for population protocols, $\mathsf{post}(\boldsymbol{c})$ is finite for all configurations $\boldsymbol{c}$, which is not true for CRNs in general. Fortunately, one may still obtain the full reverse direction of Theorem 2.8 by showing that every sym-CRD is semilinear (cf. the proof of Theorem 3.4 below) and then invoking Corollary 3.3.

We now use this machinery to re-prove the result, due originally to Angluin, Aspnes, and Eisenstat [4], that sym-CRDs decide only semilinear sets.

Theorem 3.4. *Every sym-CRD decides a semilinear set.*

Proof. Let $\mathcal{D} = (\mathcal{N}, \mathcal{I}, \mathcal{O}_0, \mathcal{O}_1)$ be a sym-CRD. Let $\mathcal{I}' = \{\boldsymbol{c} \in \mathbb{N}^\Lambda \mid \boldsymbol{c}\!\restriction_{\Lambda \setminus \Sigma} = \boldsymbol{0}\}$. The complement of $\mathcal{I}'$ is closed upwards, thus $\mathcal{I}'$ is semilinear, as is $\mathcal{I} = \mathcal{I}' \setminus \{\boldsymbol{0}\}$.

We now show that each $\mathcal{O}_i$ is semilinear. Let $\mathcal{L}_i = \{\boldsymbol{c} \in \mathbb{N}^\Lambda \mid \boldsymbol{c}\!\restriction_{\Gamma_i} \neq \boldsymbol{0}\}$ as in Definition 2.2. By Remark 2.4, $\mathcal{O}_i = \mathbb{N}^\Lambda \setminus \mathsf{pre}(\mathcal{L}_{1-i} \cup \{\boldsymbol{0}\}) = \mathbb{N}^\Lambda \setminus (\mathsf{pre}(\mathcal{L}_{1-i}) \cup \mathsf{pre}(\boldsymbol{0}))$. By Lemma 2.11 we may assume that each reaction $\alpha = (\boldsymbol{r}, \boldsymbol{p})$ of $\mathcal{D}$ has $\boldsymbol{p} \neq \boldsymbol{0}$, so $\mathsf{pre}(\boldsymbol{0}) = \{\boldsymbol{0}\}$, which is semilinear. Since $\mathcal{L}_{1-i}$ is closed upwards, by Lemma 2.10, $\mathsf{pre}(\mathcal{L}_{1-i})$ is also closed upwards, so semilinear by Lemma 2.9. Since semilinear sets are closed under union and complement, $\mathcal{O}_i$ is also semilinear, so $\mathcal{D}$ is a semilinear gen-CRD. The theorem follows by Corollary 3.3.   □

Remark 3.5. From the hypothesis $\mathsf{post}(\mathcal{I}_i) \subseteq \mathsf{pre}(\mathcal{O}_i)$ in Definition 2.2, we used only the weaker conclusion $\mathcal{I}_i = \mathcal{I} \cap \mathsf{pre}(\mathcal{O}_i)$. In other words, we need merely that $\mathcal{O}_i$ is *initially* reachable from $\mathcal{I}_i$ itself (*and* that $\mathcal{O}_{1-i}$ is unreachable from $\mathcal{I}_i$, since $\mathsf{pre}(\mathcal{O}_0)$ and $\mathsf{pre}(\mathcal{O}_1)$ partition $\mathcal{I}$). We do not require that $\mathcal{O}_i$ *remains*

---

[6] While Definition 3.1 appears almost too general to be useful, Corollary 3.3 says that if $\mathcal{I}, \mathcal{O}_0, \mathcal{O}_1$ are semilinear, then so are $\mathcal{I}_0, \mathcal{I}_1$, which implies that any CRD definition that can be framed as such a gen-CRD must decide only semilinear sets.

reachable from every configuration reachable from $\mathcal{I}_i$ (i.e., $\mathsf{post}(\mathcal{I}_i)$). Hence one could weaken part 3 of Definition 2.2 to use the condition $\mathcal{I}_i = \mathcal{I} \cap \mathsf{pre}(\mathcal{O}_i)$, and Theorem 3.4 still holds.[7]

Despite Remark 3.5, if a gen-CRD *does* obey the stronger condition $\mathsf{post}(\mathcal{I}_i) \subseteq \mathsf{pre}(\mathcal{O}_i)$, then a convenient property holds: each $\mathcal{O}_i$ may be enlarged without altering the set $\mathcal{I}_1$ decided by the gen-CRD, so long as $\mathcal{O}_{1-i}$ remains unreachable from $\mathcal{O}_i$. The following lemma formalizes this.

Lemma 3.6. *Let $\mathcal{D} = (\mathcal{N}, \mathcal{I}, \mathcal{O}_0, \mathcal{O}_1)$ be a gen-CRD that decides $\mathcal{I}_1$ and let $\mathcal{I}_0 = \mathcal{I} \setminus \mathcal{I}_1$. For $i \in \{0,1\}$, assume that $\mathsf{post}(\mathcal{I}_i) \subseteq \mathsf{pre}(\mathcal{O}_i)$, and let $\mathcal{O}'_i \supseteq \mathcal{O}_i$ with $\mathsf{post}(\mathcal{O}'_i) \cap \mathcal{O}_{1-i} = \varnothing$. Then $\mathcal{D}' = (\mathcal{N}, \mathcal{I}, \mathcal{O}'_0, \mathcal{O}'_1)$ is a gen-CRD deciding $\mathcal{I}_1$.*

Proof. We have $\mathcal{I}_i = \mathsf{pre}(\mathcal{O}_i) \cap \mathcal{I} \subseteq \mathsf{pre}(\mathcal{O}'_i) \cap \mathcal{I}$ for $i \in \{0,1\}$. To show that this inclusion is an equality, it suffices to show that $\mathsf{pre}(\mathcal{O}'_0) \cap \mathcal{I}$ and $\mathsf{pre}(\mathcal{O}'_1) \cap \mathcal{I}$ are disjoint.

Let $i \in \mathcal{I}_i$. Then $i \in \mathsf{pre}(\mathcal{O}_i) \subseteq \mathsf{pre}(\mathcal{O}'_i)$. Assume to the contrary $i \in \mathsf{pre}(\mathcal{O}'_{1-i})$. Let $o \in \mathcal{O}'_{1-i} \cap \mathsf{post}(i)$, so $o \in \mathsf{post}(i) \subseteq \mathsf{post}(\mathcal{I}_i) \subseteq \mathsf{pre}(\mathcal{O}_i)$. Thus $\mathcal{O}'_{1-i} \cap \mathsf{pre}(\mathcal{O}_i) \neq \varnothing$. In other words, $\mathsf{post}(\mathcal{O}'_{1-i}) \cap \mathcal{O}_i \neq \varnothing$ — a contradiction. Hence $\mathsf{pre}(\mathcal{O}'_0) \cap \mathcal{I}$ and $\mathsf{pre}(\mathcal{O}'_1) \cap \mathcal{I}$ are disjoint.   □

4   Asymmetric Output-Stability

We now give a natural alternative output convention for CRDs, which we call an asymmetric output-stable CRD (asym-CRD). Whereas the output i of a sym-CRD is based on both the presence of species of one type Γ_i and the absence of a species of a different type Γ_{1-i}, the output of an asym-CRD is based solely on the presence or absence of a single species type Γ_1.

For each $i \in \mathcal{I}$ the CRD can either (1) reach a configuration o so that for each configuration o' reachable from o (including o itself) we have $o' \restriction_{\Gamma_1} \neq \mathbf{0}$ or (2) reach a configuration o so that for each configuration o' reachable from o we have $o' \restriction_{\Gamma_1} = \mathbf{0}$. Similarly to gen-CRDs, and unlike sym-CRDs,[8] it is not required that such a configuration o is reachable from *any* configuration c reachable from the initial i, merely that such a o is reachable from i itself. Even this more liberal assumption does not allow the CRD to decide a non-semilinear set.

Definition 4.1. *An asymmetric output-stable chemical reaction decider (asym-CRD) is a gen-CRD $\mathcal{D} = (\mathcal{N}, \mathcal{I}, \mathcal{O}_0, \mathcal{O}_1)$, where there are $\Sigma \subseteq \Lambda$ and voting species $\Gamma_1 \subseteq \Lambda$ such that*

---

[7] In contrast, the proof of [4] crucially requires the hypothesis $\mathsf{post}(\mathcal{I}_i) \subseteq \mathsf{pre}(\mathcal{O}_i)$.

[8] As noted, sym-CRDs could be defined by replacing the requirement $\mathsf{post}(\mathcal{I}_i) \subseteq \mathsf{pre}(\mathcal{O}_i)$ with $\mathcal{I}_i = \mathcal{I} \cap \mathsf{pre}(\mathcal{O}_i)$ and retain the same power, but for clarity we retain the original definition.

1. $\mathcal{I} = \{\mathbf{c} \in \mathbb{N}^\Lambda \mid \mathbf{c}\!\restriction_{\Lambda\backslash\Sigma} = \mathbf{0}\} \setminus \{\mathbf{0}\}$, *and*
2. $\mathcal{O}_i = \{\mathbf{c} \in \mathbb{N}^\Lambda \mid \mathsf{post}(\mathbf{c}) \subseteq \mathcal{V}_i\}$ *for* $i \in \{0,1\}$, *with* $\mathcal{V}_1 = \{\mathbf{c} \in \mathbb{N}^\Lambda \mid \mathbf{c}\!\restriction_{\Gamma_1} \neq \mathbf{0}\}$ *and* $\mathcal{V}_0 = \mathbb{N}^\Lambda \setminus \mathcal{V}_1.$[9]

Condition 1 states that only species in Σ may be present initially, and at least one must be present. Condition 2 defines $\mathcal{V}_1$ and $\mathcal{V}_0$ to be configurations with and without Γ_1 voters, and $\mathcal{O}_i$ to be the stable subsets of $\mathcal{V}_i$.

Example 4.2. Consider the following asym-CRD $\mathcal{D}'$, where $\Sigma = \{X, Y\}$ and $\Gamma_1 = \{X, Y\}$, which decides the same set as in Example 2.7 (i.e., $x \not\equiv y \mod m$).

$$mX \rightarrow \varnothing, \quad mY \rightarrow \varnothing, \quad X + Y \rightarrow \varnothing. \tag{4.1}$$

If $x \equiv y \mod m$, then eventually all X and Y molecules are consumed and we obtain configuration $\mathbf{c} = \mathbf{0} \in \mathcal{O}_0$. Otherwise, all X and Y cannot be consumed, and we are in $\mathcal{O}_1$. This example illustrates that the asym-CRD computing convention may permit a simpler implementation in some cases. Indeed, compared with Example 2.7, (4.1) has 2 fewer reactions and 1 fewer species (and is also "faster" since fewer reactions need to occur).

We first observe that asym-CRDs have at least the computational power of sym-CRDs.

Observation 4.3. *Let* $\mathcal{D} = (\mathcal{N}, \mathcal{I}, \mathcal{O}_0, \mathcal{O}_1)$ *be a sym-CRD deciding* X, *with voter partition* $\{\Gamma_0, \Gamma_1\}$. *Then* $\mathcal{D}' = (\mathcal{N}, \mathcal{I}, \mathcal{O}'_0, \mathcal{O}'_1)$, *where, for* $i \in \{0,1\}$, $\mathcal{O}'_i = \{\mathbf{c} \in \mathbb{N}^\Lambda \mid \mathsf{post}(\mathbf{c}) \subseteq \mathcal{V}_i\}$, *with* $\mathcal{V}_i$ *as in Definition 4.1 (with respect to* Γ_1), *is an asym-CRD deciding* X.

Proof. This follows from Lemma 3.6 since (1) $\mathcal{O}_i \subseteq \mathcal{O}'_i$ and (2) $\mathsf{post}(\mathcal{O}'_i) = \mathcal{O}'_i$ is disjoint from $\mathcal{O}_{1-i}$ for $i \in \{0,1\}$. □

We now show that asym-CRDs have *no greater* computational power than sym-CRDs. This is not as immediate as the other direction. First, observe that an asym-CRD may not be a sym-CRD; if we interpret species $V_0 \in \Lambda \setminus \Gamma_1$ as voting "0", then a sym-CRD is required to *eliminate* them to output "1", but not an asym-CRD. Moreover, a direct transformation of an asym-CRD into a sym-CRD appears difficult. Intuitively, the problem is that the absence of molecules in Γ_1 is not detectable by a CRN, so there is no obvious way to ensure that a species $V_0 \in \Lambda \setminus \Gamma_1$ is produced only if all $V_1 \in \Gamma_1$ are absent. The next obvious proof strategy would be to show, as in the proof of Theorem 3.4, that every asym-CRD is a semilinear gen-CRD. However, it is not clear whether $\mathcal{O}_1$ is semilinear. Nonetheless, due to the generality of Definition 3.1 and Theorem 3.2, we can define a semilinear gen-CRD that decides the same set, by taking a subset of $\mathcal{O}_1$ that is provably semilinear and still satisfies the necessary reachability

---

[9] Just as for sym-CRDs, $\mathsf{post}(\mathcal{O}_i) = \mathcal{O}_i$. Note that $\mathcal{V}_1$ above is the same as $\mathcal{L}_1$ in Definition 2.2, but $\mathcal{L}_0 \neq \mathcal{V}_0$, since $\mathcal{L}_1$ and $\mathcal{L}_0$ can have nonempty intersection if there are conflicting voters present in some configuration.

constraints, even though the gen-CRD we define is *not* in fact an asym-CRD (in particular, its "output" set $\mathcal{O}_1$ is not closed under application of reactions).

Recall that a *homomorphism* $f : \mathbb{N}^\Lambda \to \mathbb{Z}$ obeys $f(c+c') = f(c)+f(c')$ for all $c, c' \in \mathbb{N}^\Lambda$. Some examples include $f(c) = c(S)$ for some $S \in \Lambda$, $f(c) = \|c\restriction_\Delta\|$ for some $\Delta \subseteq \Lambda$, or $f(c) = c(S_1) - c(S_2)$ for some $S_1, S_2 \in \Lambda$.

For a CRN $\mathcal{N}$ and a function $f : \mathbb{N}^\Lambda \to \mathbb{Z}$, we define $\mathsf{nondec}_f = \{c \in \mathbb{N}^\Lambda \mid \forall c' \in \mathsf{post}(c), f(c') \geq f(c)\}$ as the set of configurations c in which f is minimal among all the configurations reachable from c.

We now prove a key lemma, which will be used for characterizing both asym-CRDs in this section and dem-CRDs in Sect. 5.

Lemma 4.4. *Let* $\mathcal{N}$ *be a CRN and* $f : \mathbb{N}^\Lambda \to \mathbb{Z}$ *a homomorphism. Let* $\mathcal{O} = \{c \in \mathbb{N}^\Lambda \mid \mathsf{post}(c) \subseteq \mathcal{V}\}$ *with* $\mathcal{V} = \{c \in \mathbb{N}^\Lambda \mid f(c) > 0\}$. *Then* $\mathcal{O} \cap W$ *is semilinear and* $\mathsf{pre}(\mathcal{O} \cap W) = \mathsf{pre}(\mathcal{O})$, *where* $W = \mathsf{nondec}_f$.

Proof. We first prove $\mathsf{pre}(\mathcal{O} \cap W) = \mathsf{pre}(\mathcal{O})$. Obviously, $\mathsf{pre}(\mathcal{O} \cap W) \subseteq \mathsf{pre}(\mathcal{O})$. To prove the reverse containment, let $c \in \mathsf{pre}(\mathcal{O})$. Hence $c \in \mathsf{pre}(o)$ for some $o \in \mathcal{O}$. Since every $o' \in \mathsf{post}(o)$ satisfies $f(o') > 0$, there is $o' \in \mathsf{post}(o)$ such that $f(o')$ is minimal among all configurations in $\mathsf{post}(o)$. Thus $o' \in W$. Since $\mathsf{post}(\mathcal{O}) = \mathcal{O}$, we have $o' \in \mathcal{O}$. Hence, $o' \in \mathcal{O} \cap W$. Now, $o \in \mathsf{pre}(o')$ and $c \in \mathsf{pre}(o)$, and so $c \in \mathsf{pre}(o')$. Therefore, $c \in \mathsf{pre}(\mathcal{O} \cap W)$, so $\mathsf{pre}(\mathcal{O}) \subseteq \mathsf{pre}(\mathcal{O} \cap W)$.

We now show that $\mathcal{O} \cap W$ is semilinear. Observe that the set $\mathbb{N}^\Lambda \setminus W = \{c \in \mathbb{N}^\Lambda \mid \exists c' \in \mathsf{post}(c), f(c') < f(c)\}$ is closed upwards. Indeed, if $c \in \mathbb{N}^\Lambda \setminus W$ and $c' \in \mathsf{post}(c)$ with $f(c') < f(c)$, then for all $d \in \mathbb{N}^\Lambda$, $c' + d \in \mathsf{post}(c + d)$ and $f(c' + d) = f(c') + f(d) < f(c) + f(d) = f(c + d)$. Thus $\mathbb{N}^\Lambda \setminus W$ is semilinear by Lemma 2.9, and hence also W. Since $\mathcal{O} \subseteq \mathcal{V}$, we have $\mathcal{O} \cap W \subseteq \mathcal{V} \cap W$. Conversely, if $c \in \mathcal{V} \cap W$, then $f(c) > 0$ since $c \in \mathcal{V}$, and for all $c' \in \mathsf{post}(c)$, $f(c') \geq f(c) > 0$ since $c \in W$. Thus $c \in \mathcal{O} \cap W$, showing $\mathcal{O} \cap W = \mathcal{V} \cap W$, which is semilinear since $\mathcal{V}$ and W are.   □

Using Lemma 4.4 we show that every asym-CRD can be changed into a semilinear gen-CRD by choosing $\mathcal{O}_1 \cap W$, rather than $\mathcal{O}_1$, as its "output 1" set of configurations. Note that unlike in the definition of sym-CRD and asym-CRD, $\mathcal{O}_1 \cap W$ is *not* in general closed under application of reactions.

Lemma 4.5. *Let* $\mathcal{D} = (\mathcal{N}, \mathcal{I}, \mathcal{O}_0, \mathcal{O}_1)$ *be an asym-CRD deciding* X *and* Γ_1 *be as in Definition 4.1. Let* $W = \mathsf{nondec}_f(\Gamma_1)$ *with* $f : \mathbb{N}^\Lambda \to \mathbb{Z}$ *defined as* $f(c) = \|c\restriction_{\Gamma_1}\|$ *for all* $c \in \mathbb{N}^\Lambda$. *Then* $\mathcal{D}' = (\mathcal{N}, \mathcal{I}, \mathcal{O}_0, \mathcal{O}_1 \cap W)$ *is a semilinear gen-CRD deciding* X.

Proof. Observe that f is a homomorphism. Now, Lemma 4.4 tells us that $\mathsf{pre}(\mathcal{O}_1 \cap W) = \mathsf{pre}(\mathcal{O}_1)$; thus $\mathcal{D}'$ decides X.

To complete the proof, it suffices to show that $\mathcal{D}'$ is semilinear. $\mathcal{I}$ is obtained from the closed-upwards set $\mathbb{N}^\Sigma \setminus \{0\}$ by padding zeros for the species of $\Lambda \setminus \Sigma$, so $\mathcal{I}$ is semilinear. $\mathcal{O}_1 \cap W$ is semilinear by Lemma 4.4. To see that $\mathcal{O}_0$ is semilinear, let $\mathcal{V}_0$ and $\mathcal{V}_1$ be as in Definition 4.1. Clearly $\mathcal{V}_1$ is closed upwards, so semilinear. So, (1) $\mathsf{pre}(\mathcal{V}_1)$ is also closed upwards and therefore semilinear (by Lemmas 2.9 and 2.10) and (2) $\mathcal{V}_0 = \mathbb{N}^\Lambda \setminus \mathcal{V}_1$ is semilinear. Thus, $\mathcal{O}_0 = \mathcal{V}_0 \setminus \mathsf{pre}(\mathcal{V}_1)$ is semilinear since the class of semilinear sets is closed under set difference.   □

The following is the first of two main results of this paper. It says that the computational power of sym-CRDs equals that of asym-CRDs; they both decide exactly the semilinear sets.

Theorem 4.6. *Let $X \subseteq \mathbb{N}^\Sigma \setminus \{0\}$. Then X is semilinear if and only if there is an asym-CRD that decides X.*

Proof. The forward direction follows from Observation 4.3 and Theorem 2.8. For the reverse direction, let $\mathcal{D}$ be an asym-CRD deciding X. By Lemma 4.5, there is a semilinear gen-CRD $\mathcal{D}'$ deciding X, which is semilinear by Corollary 3.3. □

5   Democratic Output-Stability

Another reasonable alternative output convention is the one most naturally associated with the term "voting": a *democratic* output convention in which, rather than requiring a consensus, we define output by majority vote. In this case, for sets of voting species Γ_0 and Γ_1, the only undefined outputs occur in "tie" configurations c where $\|c{\restriction}_{\Gamma_0}\| = \|c{\restriction}_{\Gamma_1}\|$. In this section we show that such CRDs have equivalent computing power to sym-CRDs.

Definition 5.1. *A democratic output-stable chemical reaction decider (dem-CRD) is a gen-CRD $\mathcal{D} = (\mathcal{N}, \mathcal{I}, \mathcal{O}_0, \mathcal{O}_1)$, where there are $\Sigma \subseteq \Lambda$ and a partition $\{\Gamma_0, \Gamma_1\}$ of Λ such that*

1. $\mathcal{I} = \{c \in \mathbb{N}^\Lambda \mid c{\restriction}_{\Lambda \setminus \Sigma} = 0\} \setminus \{0\}$,
2. $\mathcal{O}_i = \{c \in \mathbb{N}^\Lambda \mid \mathsf{post}(c) \subseteq \mathcal{M}_i\}$, *with* $\mathcal{M}_i = \{c \in \mathbb{N}^\Lambda \mid \|c{\restriction}_{\Gamma_i}\| > \|c{\restriction}_{\Gamma_{1-i}}\|\}$ *for* $i \in \{0, 1\}$.

Note that $\mathcal{M}_0 \cap \mathcal{M}_1 = \varnothing$, and that $\mathcal{O}_i$ is stable, i.e., $\mathcal{O}_i = \mathsf{post}(\mathcal{O}_i)$. A sym-CRD reaches a consensus, the strongest kind of majority, leading to the following observation implying that dem-CRDs are at least as powerful as sym-CRDs.

Observation 5.2. *Let $\mathcal{D} = (\mathcal{N}, \mathcal{I}, \mathcal{O}_0, \mathcal{O}_1)$ be a sym-CRD deciding X, with voter partition $\{\Gamma_0, \Gamma_1\}$. Then $\mathcal{D}' = (\mathcal{N}, \mathcal{I}, \mathcal{O}_0', \mathcal{O}_1')$, where $\mathcal{O}_i' = \{c \in \mathbb{N}^\Lambda \mid \mathsf{post}(c) \subseteq \mathcal{M}_i\}$ for $i \in \{0, 1\}$, with $\mathcal{M}_i$ as in Definition 5.1, is a dem-CRD deciding X.*

Proof. This follows from Lemma 3.6 since (1) $\mathcal{O}_i \subseteq \mathcal{O}_i'$ and (2) $\mathsf{post}(\mathcal{O}_i') = \mathcal{O}_i'$ is disjoint from $\mathcal{O}_{1-i}$ for $i \in \{0, 1\}$.   □

The converse result, that dem-CRDs are no more powerful than sym-CRDs, implies the second main result of this paper. The proof of the following theorem is found in the full version of this paper, and relies on the gen-CRD framework of Sect. 3 and Lemma 4.4 (choosing f that is the difference between 0 and 1 voters).

Theorem 5.3. *Let $X \subseteq \mathbb{N}^\Sigma \setminus \{0\}$. Then X is semilinear if and only if there is a dem-CRD that decides X.*

6   Discussion

Using a recent result about Petri nets [13] (cf. Theorem 3.2) we have presented a framework able to capture different output conventions for computational CRNs. The original symmetric consensus-based definition [3] can be fitted in this framework, giving a new proof that such CRNs are limited to computing only semilinear sets. Two additional definitions, an asymmetric existence-based convention, and a symmetric majority-vote convention, can be fitted in this framework, and thus have the same expressive power as the original.

We show that asym-CRDs and dem-CRDs are no more powerful than sym-CRDs by showing that they are limited to deciding semilinear sets, which is known also to apply to sym-CRDs. It would be informative, however, to find a proof that uses a direct simulation argument, showing how to transform an arbitrary asym-CRD or dem-CRD into a sym-CRD deciding the same set. Along a similar line of thinking, we have defined the computational ability of CRDs without regard to time complexity, which is potentially sensitive to definitional choices, even if the class of decidable sets remains the same [1,2,5,11,12]. It would be interesting to find cases in which asym-CRDs or dem-CRDs are be able to compute faster than any equivalent sym-CRD.

An open problem is to consider other output conventions, where we possibly step out of semilinearity. For example, consider a designated species V_1 such that for each input configuration $d \in \mathcal{I}$, (1) $d \in \mathcal{I}_1$ if we always eventually reach a configuration c such that all configurations reachable from c has a V_1 molecule, and (2) $d \in \mathcal{I}_0$ if we can never reach such a configuration c. Hence the output of a configuration is then based on a behavioral property of the system (whether it is stable) instead of a syntactic property of the configuration (whether it contains a particular molecule). It is not clear how to apply Theorem 3.2, which requires that $\mathcal{I}_0 = \mathcal{I} \cap \mathsf{pre}(S)$ for some semilinear set S.

Acknowledgements. R.B. thanks Grzegorz Rozenberg for interesting and useful discussions regarding chemical reaction networks. D.D. thanks Ryan James for suggesting the democratic CRD model. The authors are grateful to the anonymous reviewers for comments that have helped improve the presentation.

References

1. Alistarh, D., Aspnes, J., Eisenstat, D., Gelashvili, R., Rivest, R.L.: Time-space trade-offs in population protocols. Technical report (2016). arXiv: 1602.08032
2. Alistarh, D., Gelashvili, R.: Polylogarithmic-time leader election in population protocols. In: Halldórsson, M.M., Iwama, K., Kobayashi, N., Speckmann, B. (eds.) ICALP 2015. LNCS, vol. 9135, pp. 479–491. Springer, Heidelberg (2015)
3. Angluin, D., Aspnes, J., Diamadi, Z., Fischer, M.J., Peralta, R.: Computation in networks of passively mobile finite-state sensors. Distrib. Comput. **18**(4), 235–253 (2006)
4. Angluin, D., Aspnes, J., Eisenstat, D.: Stably computable predicates are semilinear. In: Proceedings of the Twenty-Fifth Annual ACM Symposium on Principles of Distributed Computing, PODC 2006, pp. 292–299. ACM Press, New York (2006)

5. Angluin, D., Aspnes, J., Eisenstat, D.: Fast computation by population protocols with a leader. Distrib. Comput. **21**(3), 183–199 (2008)
6. Angluin, D., Aspnes, J., Eisenstat, D., Ruppert, E.: The computational power of population protocols. Distrib. Comput. **20**(4), 279–304 (2007)
7. Brijder, R.: Output Stability and semilinear sets in chemical reaction networks and deciders. In: Murata, S., Kobayashi, S. (eds.) DNA 2014. LNCS, vol. 8727, pp. 100–113. Springer, Heidelberg (2014)
8. Chen, H.-L., Doty, D., Soloveichik, D.: Deterministic function computation with chemical reaction networks. Nat. Comput. **13**(4), 517–534 (2014)
9. Cummings, R., Doty, D., Soloveichik, D.: Probability 1 computation with chemical reaction networks. Nat. Comput. **15**, 1–17 (2015)
10. Dickson, L.E.: Finiteness of the odd perfect and primitive abundant numbers with n distinct prime factors. Am. J. Math. **35**, 413–422 (1913)
11. Doty, D., Hajiaghayi, M.: Leaderless deterministic chemical reaction networks. Nat. Comput. **14**(2), 213–223 (2015)
12. Doty, D., Soloveichik, D.: Stable leader election in population protocols requires linear time. In: Moses, Y. (ed.) DISC 2015. LNCS, vol. 9363, pp. 602–616. Springer, Heidelberg (2015)
13. Esparza, J., Ganty, P., Leroux, J., Majumdar, R.: Verification of population protocols. In 26th International Conference on Concurrency Theory (CONCUR 2015), vol. 42, pp. 470–482 (2015)
14. Gillespie, D.T.: Exact stochastic simulation of coupled chemical reactions. J. Phys. Chem. **81**(25), 2340–2361 (1977)
15. Ginsburg, S., Spanier, E.H.: Semigroups, Presburger formulas, and languages. Pac. J. Math. **16**(2), 285–296 (1966)
16. Hopcroft, J.E., Pansiot, J.-J.: On the reachability problem for 5-dimensional vector addition systems. Theoret. Comput. Sci. **8**, 135–159 (1979)
17. Karp, R.M., Miller, R.E.: Parallel program schemata. J. Comput. Syst. Sci. **3**(2), 147–195 (1969)
18. Mealy, G.H.: A method for synthesizing sequential circuits. Bell Syst. Tech. J. **34**(5), 1045–1079 (1955)
19. Moore, E.F.: Gedanken-experiments on sequential machines. Automata Stud. **34**, 129–153 (1956)
20. Peterson, J.L.: Petri nets. ACM Comput. Surv. **9**(3), 223–252 (1977)
21. Soloveichik, D., Cook, M., Winfree, E., Bruck, J.: Computation with finite stochastic chemical reaction networks. Nat. Comput. **7**(4), 615–633 (2008)
22. Soloveichik, D., Seelig, G., Winfree, E.: DNA as a universal substrate for chemical kinetics. Proc. Natl. Acad. Sci. **107**(12), 5393–5398 (2010)

Chemical Reaction Network Designs
for Asynchronous Logic Circuits

Luca Cardelli[1,2], Marta Kwiatkowska[1], and Max Whitby[1(✉)]

[1] Department of Computer Science, University of Oxford, Oxford, UK
[2] Microsoft Research, Cambridge, UK
luca@microsoft.com, max.whitby@keble.ox.ac.uk

Abstract. Chemical reaction networks (CRNs) are a versatile language for describing the dynamical behaviour of chemical kinetics, capable of modelling a variety of digital and analogue processes. While CRN designs for synchronous sequential logic circuits have been proposed and their implementation in DNA demonstrated, a physical realisation of these devices is difficult because of their reliance on a clock. Asynchronous sequential logic, on the other hand, does not require a clock, and instead relies on handshaking protocols to ensure the temporal ordering of different phases of the computation. This paper provides novel CRN designs for the construction of asynchronous logic, arithmetic and control flow elements based on a bi-molecular reaction motif with uniform reaction rates. We model and validate the designs using Microsoft's GEC tool.

1 Introduction

Chemical Reaction Networks (CRNs) are traditionally used to capture the behaviour of inorganic and organic chemical reactions in a well-mixed solution. Recently, a paradigm shift in the scientific community has seen the use of CRNs extend to that of a high-level programming language for molecular computing devices [10], where the fundamental computational process differs from conventional digital electronics in that it involves transformation of input chemicals into output via reaction rules. Several digital and analogue circuits [17,25] have been designed in CRNs and their computational power studied [7,26]. It has also been demonstrated in principle that any CRN can be physically realised in DNA [3,9,26]. CRNs are therefore particularly attractive as a programming language for use in nanotechnology and biomedical applications, where it is difficult to integrate traditional electronics.

While CRN designs for synchronous sequential logic circuits have been proposed, a physical realisation of these devices is challenging because of their reliance on a clock to synchronise events in order to ensure the correct temporal order of the phases of the computation. Clocks are difficult to make, since they arise from unique conditions of chemical concentrations and kinetic constants,

This research is supported by a Royal Society Research Professorship and ERC AdG VERIWARE.

Y. Rondelez and D. Woods (Eds.): DNA 2016, LNCS 9818, pp. 67–81, 2016.
DOI: 10.1007/978-3-319-43994-5_5

and must control a large number of events. In electronics, an alternative circuit design technology is asynchronous sequential logic [27], which instead of a clock relies on handshaking protocols to synchronise events. Asynchronous circuits are widely used for low-power microprocessor designs, e.g., by ARM, though require a larger circuit area. The key component is the Muller C-element, which is used to synchronise multiple independent processes. To ensure Turing completeness of asynchronous circuits, we also require an isochronous fork in addition to the Muller C-element. An isochronous fork is a component which produces a fan-out of signals that reach the target at virtually the same time. This assumption is difficult to achieve in conventional electronics, because of the need to make the wires the same length, but is straightforward in chemical kinetics because of the well-mixed assumption.

This paper provides novel CRN designs for the construction of an asynchronous computing device based on a bi-molecular reaction motif inspired by the Approximate Majority network [1,5]. All components are produced with simple reactions and uniform reaction rates, and are independent of a universal clock. Moreover, any design provided in this paper could in principle be realised as a two-domain DNA strand displacement device [3].

We work with the dual-rail design methodology and employ a variant of the diagrammatic language of [4] to represent the designs at the high level. Starting from the Muller C-element, we design the main components of a complete asynchronous computing device in terms of CRNs in a principled way, including logic gates, control flow and basic arithmetic. We illustrate the designs on selected components validated using Microsoft's Visual GEC tool[1], both for the deterministic and stochastic semantics, with the latter approximated using a prototype implementation of the Linear Noise Approximation of [6].

Our designs constitute the first feasible implementation of asynchronous computational components as CRNs, and are relevant for a multitude of applications in synthetic biology and biosensing.

2   Related Work

The computational power of CRNs, viewed as a programming language for engineering biochemical systems, has been studied by a number of authors, to mention [7,10]. Researchers have investigated their power to simulate Boolean circuits, molecular machines, or distributed algorithms [1,11,17,25,26]. Assuming a small probability of error, CRNs have been shown to be Turing-universal [25]. Since the behaviour of CRNs is asynchronous, a fact evident through their equivalence with Petri net models [10], the main difficulty with programming them is the need to control the order of reactions. In [10] it is suggested that this "uncontrollability" can be handled by changing rate constants, an idea followed up in [19], where CRN designs for basic arithmetic are given based on two rate constants, "fast" and "slow". Our designs, on the other hand, exploit the asynchrony of the underlying CRN model and work with uniform rates.

---

[1] http://lepton.research.microsoft.com/webgec/.

In [10] we see the construction and composition of simple logic gates based upon catalytic reactions, but they do not mention control flow or systematic component design in a dual rail setting. In [23] the authors propose CRNs for an inverter, an incrementer, a decrementer, and a copier; their designs are based on two rate constants, "fast" and "slow", and thus are not rate-independent. A system of actual chemical reactions is found in [12], where a precise molecular implementation is given for gates complete with a thermodynamic analysis of how the system would evolve, though only for simple gate designs. An implementation of individual dual-rail logic gates that are rate-independent is given in [8]. In contrast, our designs are composable and capable of performing non-trivial computation.

Designs for the Muller C-element, though not the remaining components of an asynchronous device, have been constructed from genetic logic gates [20] and a genetic toggle switch [21], but we are not aware of any other nanoscale designs for asynchronous circuits.

3   Preliminaries

3.1   Chemical Reaction Networks

A CRN C is a finite set R of reactions acting on a finite set S of species. A reaction, which can either be reversible or irreversible, is a triple written in the form $\langle r \in S^{\mathbb{N}}, k \in \mathbb{R}_{>0}, p \in S^{\mathbb{N}} \rangle$, where r and p are the multisets of species reactants and products, respectively, and $k > 0$ is the reaction rate [25]. We work with bi-molecular reactions with uniform rates, including catalytic reactions, and assume mass-action kinetics.

The *stochastic* semantics of a CRN [13] can be given as a continuous-time Markov chain with the state space given as discrete vectors of population counts, which can be solved through the Chemical Master Equation (CME) whose numerical solution may be infeasible for large molecular counts. The *deterministic* semantics approximates the species concentrations over time as a solution of rate equations [28], assuming a continuous state space, but is valid only for high molecular counts and cannot model stochastic fluctuations. A *stochastic approximation* of the CME is possible using the Linear Noise Approximation (LNA) [28], which provides Gaussian distributions for variance. This is a continuous approximation, which is independent of initial populations and hence scalable, and is valid in the limit of high populations. LNA was recently adapted to provide stochastic analysis of the evolution of populations of molecular species of CRNs [6] and extended to probabilistic reachability in [2].

We emphasise that we work with CRNs as an *abstract* programming language for artificial devices, as argued in [10], where we assume that molecules can be designed to carry out the required reactions. In principle, any CRN can be implemented using nucleic-acid strand displacement cascades [3,26], which has been recently experimentally demonstrated in [9].

3.2   Principles of Asynchronous Circuit Design

Asynchronous computation [27] is Turing complete [18] meaning that any bounded-tape Turing machine can be implemented with an asynchronous circuit, providing that the implementation of that circuit has isochronous forks. An isochronous fork is the propagation of a signal from a single source to multiple receivers with the important constraint that the signal must reach the receivers at precisely the same time. In classical digital circuitry this could be seen as the propagation of a signal down wires of exactly the same length from one component to another.

Asynchronous computation relies on 'local cooperation' in the form of handshaking protocols, rather than a governing clock. These protocols exchange completion signals (high or low, also denoted 1 or 0) in order to establish when a computation has terminated. Asynchronous circuits rely heavily on latches and rendez-vous elements. A rendez-vous element is a component which 'waits' on two or more actions to complete before a system continues. One form of a rendez-vous element is the Muller C-element [27, p. 5], which has two Boolean inputs and one output, and is "stateful". When both inputs are the same, the output switches, if necessary, to be equal to the inputs, but when the inputs are different the output remains what it was last time the inputs were equal. The C-element suffices to build a gate that synchronises events, but an isochronous fork, which produces a fan-out of signals that reach the target at virtually the same time, is needed to ensure Turing completeness [18].

C-elements allow a circuit to be speed-independent by a series of local handshakes. This means that we can wait for longer computational paths to complete before advancing without additional computation occurring, negating the use of a system clock. A fundamental construct built from C-elements is a Muller pipeline, shown in Fig. 1, which is used to relay handshakes and can be combined with data storage or computational components. The Muller pipeline is constructed by the composition of Muller C-elements and NOT-gates. Initially all C-elements are set to a value of 0. The ith C-element $C[i]$ will propagate a 1 from its predecessor, $C[i-1]$, only if its successor, $C[i+1]$, is 0. Similarly, it will propagate a 0 from its predecessor only if its successor is 1. Eventually the first request initialized on the left hand side of our pipeline is propagated to the final request on the right. The protocol enacted upon this pipeline uses request and acknowledge rails that can be set to high or low. The Muller pipeline implements a basic four phase protocol, which is as follows. Firstly, the sender sends data and sets request to high, viewed in Fig. 1 as the signal $ReqHi$. The receiver then records this data and sets acknowledge to high ($AckHi$). Then the sender responds by setting request to low ($ReqLo$), and finally the receiver acknowledges this by setting acknowledgement to low ($AckLo$). If at any point a handshake along the pipeline is slower than another, the pipeline will behave like a FIFO queue with data preserved. Herein lies the important purpose of the pipeline: it allows for the delay-insensitive transfer of information from one place to another. In combination with a latch we can create the propagation of information across latches using the pipeline as a control structure.

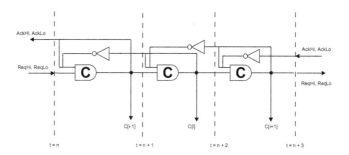

Fig. 1. Signals are propagated from left to right using a Muller pipeline. The pipeline effectively queues data, only allowing a transition to occur when a further signal has been acknowledged.

4   Circuit Construction and Design

Asynchronous computation is well suited to CRNs: they are inherently asynchronous as each reaction happens stochastically and there is no inherent clock that governs their operation.

We use a dual-rail implementation of asynchronous circuits due to the fact that we cannot detect when there are no molecules of a molecular species; we can only detect their presence. This means that there is a separate species or 'signal' representing the values high and low (or logical 1 and 0). These species are named accordingly; for instance, a signal x will be represented by two species xhi and xlo representing the high and low signals. Circuit design follows the normal rules in which components are connected by rails that transport data around the system. We assume standard knowledge of logic gates.

The designs are presented in a diagrammatic notation which allows us to view CRNs as circuits instead of a list of reactions. All our designs are built from a simple motif, seen here, which describes the two reactions:

$$xhi \xrightarrow{} \begin{array}{c} yhi \\ \updownarrow \\ ylo \end{array} \xrightarrow{} xlo$$

or in standard CRN notation:

$$xhi + ylo \xrightarrow{k} xhi + yhi \qquad xlo + yhi \xrightarrow{k} xlo + ylo \qquad (1)$$

where xhi, xlo are catalytic to the reaction $yhi \leftrightarrows ylo$ and k is the rate which is the same for both reactions, where we set $k = 1$. In general, a black circle represents that the species connected to it is catalytic to the reaction adjacent to the black circle. In our implementation, the presence of every signal is catalytic to another signal, and thus the total number of molecules for each signal (pair of species) is preserved. Implementations of catalytic gates typically rely on fuel molecules to provide energy to catalyse one species into another [26].

Our motif can be used both as a basic building block for a logical element as well as a control element. For example xhi, xlo, above, can be used to control two separate sub-circuits through yhi, ylo. This demonstrates that, through just two reactions, we can create circuits that exhibit complex behaviour both in control flow and logic.

We assume well-mixed solution, which ensures that the probability of collision between molecules is independent of their position. This yields a circuit that operates correctly but with unknown delays, called delay-insensitive [27].

4.1 Latches and the Muller C-Element

A latch is a device used in electronics to store a logical 0 or 1; it needs to have at least two stable states which are cycled between. We present three latch designs in Fig. 2(a), each intended to interface in a specific way when used within a larger system. The first, shown in Fig. 2(a-i), is almost identical to our motif except for two additional reactions which catalyse ylo to yhi, and vice versa. The latch in Fig. 2(a-ii) has an input rhi used to reset the latch to a central state. The advantage of this central state, $ymid$, is that the latch can be in a state where neither yhi nor ylo are present, which is useful if these reactions are catalytic to any other component. In its electronic counterpart, a system may have a rail deciding whether the component is active or inactive; our intermediary $ymid$ fulfils this function, as the system is in a state where neither yhi nor ylo are present. In Fig. 2(a-iii), we see the latter latch combined with control species chi, clo. These species are used to synchronise the latch in the pipeline. Even when the input signal xhi, xlo is present, we still need the species chi to be present in order to catalyse the reaction $s_2 \rightarrow shi$ or $s_4 \rightarrow slo$, which are the output species. This is needed since these output species cannot be read until the system has synchronised.

A C-element is conceptually similar to a latch except for having two inputs, x and y. The design for this, presented in Fig. 2(b-i), was inspired by the Approximate Majority (AM) circuit in [5]. The circuit in Fig. 2(b-ii) is similar to AM of [5] except for separating the inputs. Note that the C-element includes two separate AM circuits, shown schematically in Fig. 2(b-iii), where the arrows in the AM box indicate the direction of switching between the two stable states. Using this mechanism it is possible to trap the gate into the state zdn or zup given both high inputs xhi, yhi or both low inputs xlo, ylo. However, when one of these inputs is changed, we see that the gate is trapped in one of these states in view of the feedback loops. The second AM mechanism corrects the weak output signal zdn, zup from the first AM mechanism, meaning the outputs of this gate are zlo and zhi.

4.2 Logic and Arithmetic

Although gate designs for Boolean operators have been proposed in CRNs [25], we present dual-rail implementations of logic gates in line with other designs proposed within this paper. In contrast to the gates in [25], our gates account

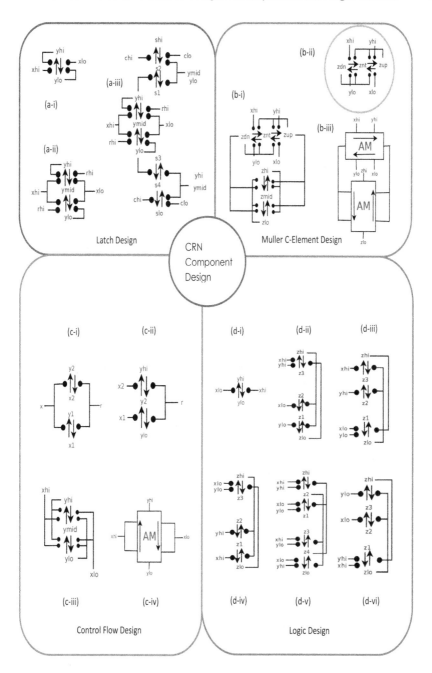

Fig. 2. a) Latches-(i) Two state latch (ii) Three state latch (iii) Latch with control input b) Muller C-Element-(i) C-Element (ii) Approximate Majority circuit based on [3] (iii) C-Element as composition of Approximate Majority circuits c) Control Flow Design-(i) Fork (ii) Join (iii) Arbiter (iv) Arbiter as AM circuit d) Logic Gates-(i) NOT (ii) AND (iii) OR (iv) NAND (v) XOR (vi) NOR

for all inputs x and y, and also respond to change in input. The AND-gate, shown in Fig. 2(d-ii), has inputs x, y expressed in our dual-rail implementation. With the presence of species yhi we can catalyse z into the state $zmid$, and with the species xhi we can catalyse $zmid$ to zhi; thus both species are needed for the gate to output the signal z. The state zhi converts any species zlo back to $z1$, therefore showing that only one output signal can be present at any time. Conversely, with either xlo, ylo we can convert $z1$ to zlo, which in turn can convert zhi back to $zmid$ and $zmid$ to $z2$. Using a similar trail of thought we can see how the other gates are devised, albeit XOR is slightly different. XOR, traditionally a gate that requires a composition of many other logic gates, has to be constructed with all combinations of inputs considered.

Using these designs, we have also implemented a ripple carry adder, seen in Fig. 7. An individual adder is composed of two XOR gates, two OR gates and an AND gate. It takes two inputs x, y and outputs the sum of the inputs z with a carry bit c. In our ripple carry implementation we compose three of these adders in series.

4.3 Control Flow

Control flow is used to mediate or propagate the flow of information throughout a system. The fork, shown Fig. 2(c-i), is used to split signals; it is constructed by having one input species x catalyse the two reactions $x_1 \rightarrow y_1$ and $x_2 \rightarrow y_2$, to produce two outputs y_1, y_2. The species r acts as a reset for the fork if the process needs repeating, assuming x is no longer active. The join, see Fig. 2 (c-ii), is similar to the function of an AND-gate, and will only output a signal yhi when both inputs x_1, x_2 are present. This is a useful control mechanism since the system can stall the catalysis of further reactions via y until both input signals x_1 and x_2 are present. The species r can be used to reset the join. We also present our AM circuit as an arbiter seen in Fig. 2(c-iii). An arbiter is used to decide an output signal based on which species arrived first. The AM circuit works well as an arbiter due to the fact that the output yhi, ylo starts to be converted from $ymid$ as soon as either of xhi, xlo arrives, therefore automatically biasing whichever species is present first. All three of these control flow elements are used in our queue and adder implementations discussed within the next section.

5 Design Validation

We use Microsoft's Visual GEC tool to establish that the designs[2] exhibit correct behaviour, both for the deterministic and stochastic semantics of the CRNs. Visual GEC provides a programming language, LBS, for designing and simulating any given CRN. Using numerical simulation, we systematically tested each component in isolation by simulating its behaviour over all inputs, and then checking that those inputs yield the desired output and also suppress any

---

[2] Available from https://github.com/max1s/CRNcode.

unwanted outputs. Next, we examined how a component might behave in a larger system, where it will be exposed to a change in input. To this end, we introduced new reactions to emulate a signal change. For instance, if we wished to change a carrier signal from high to low, we would introduce an additional reaction $xhi \xrightarrow{k} xlo$, which converts all of the signal xhi into a signal xlo while the component is operating.

Since deterministic semantics is not accurate for low molecular populations, we additionally explored its stochastic semantics. Visual GEC exports models to the probabilistic model checker PRISM [15], which then enables verification of the induced continuous-time Markov chain against temporal logic properties. This allows one to check that the circuits ensure the correct temporal ordering of the events, for example, for the Muller pipeline of Fig. 1, that the species in the first stage of the pipeline is present before the species in the second, i.e. with probability 1, and that the signal is eventually propagated to the end of the pipeline. PRISM implements numerical solution of the CME, which is exponential in the initial number of molecules and hence not scalable, and analysis based on stochastic simulation, which is time consuming. We thus used an experimental implementation of the LNA within Visual GEC, based on [6]. The LNA approximates the CME with a set of differential equations, quadratic in the number of species and independent of the initial number of molecules. The ODEs describe the time evolution of expected value and variance. As well as being capable of checking temporal logic properties [2,6], the LNA can plot the species concentration over time together with standard deviation, and is fast and reasonably accurate even for low molecule counts. Moreover, compared to the deterministic semantics, LNA provides important information about stochasticity that may affect the robustness of the circuits, and which can be explored further with CME, stochastic simulation, or verifying that the circuit converges with probability 1 to a single value.

We now illustrate the results of the validation on a selection of components.

Muller C-element. Firstly, we demonstrate the robustness of the Muller C-element against changes in input signal in Fig. 3(a-c). In Fig. 3(d), we show that the C-element may not be robust at low molecular counts, here 10. Increasing this count to 500 greatly decreases the variance of the output species (not shown), reducing the likelihood that the wrong species will reach the threshold.

Pipeline. The pipeline, seen in Fig. 1, is a mechanism that relays handshakes between components, for example latches to store data. We construct the pipeline by placing three of our C-element CRNs in sequence. At each intermediate stage between the C-elements we add a fork. One path of the fork is negated and fed back into the previous C-element, and the other path is fed into the new C-element.

Because we have already validated the individual C-element design, we can assume that they work correctly and so we only need to observe the behaviour of the overall system. We therefore analyse the system behaviour over time,

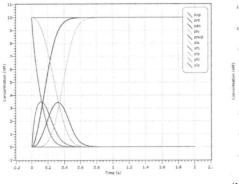

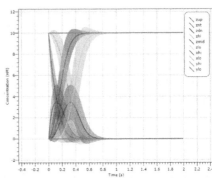

(a) An input change from xhi, yhi to xlo, ylo

(b) LNA simulation of a change in input from xhi, yhi to xlo, ylo

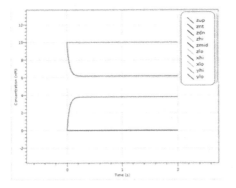

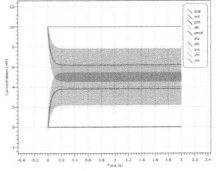

(c) An input change from xhi, yhi to xlo, yhi

(d) LNA simulation of a change in input from xhi, yhi to xlo, yhi

Fig. 3. Validation of the Muller C-element. In these experiments we start with an input of x and y, the presence of which are represented by the species xhi and yhi. In (a) we show a change of input where both x and y change to 0 or are not present, represented by species xlo and ylo. Note how zhi responds by reaching zero molecules and zlo reaches the maximum molecular value, in this case 10. In (b) we show the LNA approximation of the same scenario, with standard deviation shown as highlighted regions, which demonstrates that the variance is low once the circuit reaches steady state. In (c) we demonstrate the change in one input value xhi across a single Approximate Majority circuit; in this case the output signal decreases, but still remains at a value greater than zlo. However, for this reason we add an additional AM circuit to further separate the output signals zhi, zlo. In (d), we show the LNA of the same scenario, demonstrating that the circuit is not robust under low molecular count.

based upon a change in inputs, namely, signals $reqhi$, $reqlo$, $acchi$ and $acclo$. We conducted multiple experiments in which we change these inputs, demonstrating the desired effect of them being propagated along the pipeline. This is seen as a 'wave' through the pipeline propagating a high signal and then a low signal. The results of this are shown in Figure 4. Here the presence of the species ahi, bhi, chi represents a high signal before responding and diminishing back to zero.

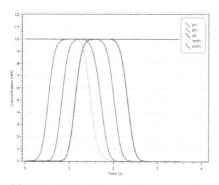 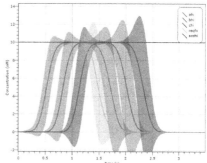

(a) The Muller C-pipeline responding to the input species *reqhi* being present and then transforming to the input species *reqlo*.

(b) The same experiment calculated with the LNA. The standard deviation is shown with highlighted regions.

Fig. 4. Validation of the Muller C-pipeline. The input request signal, encoded by the species *reqhi*, is propagated to the end of the pipeline (represented by the species *ahi, bhi, chi*); we then set the request signal to low. The pipeline then responds by the presence of *ahi, bhi* and *chi* diminishing to zero. In (b) we show that the variance is low, even for low molecular counts.

Queue. We have also designed and validated a queue, shown in Fig. 5, built by the addition of latches at each C-element block to the Muller pipeline. The queue uses the pipeline as a control mechanism to propagate signals between the latches. We use the complex latch in Fig. 2(a-iii) for this purpose. As a high species is propagated along the pipeline, it sends a signal to the queue to read and store the value in the next latch along. Each latch represents some computation that could be completed within each time interval. In Fig. 6 we analyse the oscillatory behaviour of the queue using the LNA, demonstrating its robustness at high molecular counts.

Adder. We have also designed a three bit ripple carry adder seen in Fig. 7, which works in a similar fashion to the queue but instead of latches we compose adders in series. At each time step we input two bits and a carry, which outputs the sum and a carry. In this way we can add two three-bit numbers together. We show in Fig. 8 that the adder exhibits correct behaviour, and each sum is calculated only in the next stage in the pipeline.

5.1   Discussion

Direct chemical implementations of CRNs have been theorised and realised, but involve complicated reaction mechanisms [24]. For instance, [14] implements chemical systems as neural networks. Most implementations need some external fuel molecules, as reactions such as $A + B \rightarrow C + B$ require some energy input in order to catalyse one species to another [26]. CRNs have been implemented in

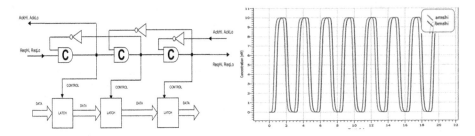

Fig. 5. Deterministic simulation of the queue pipeline. We propagate a value of 1 through the queue. The species *amshi, bmshi* represent the outputs of the first and second latches. Note that through oscillatory patterns generated by the pipeline we can mimick properties of a synchronous system.

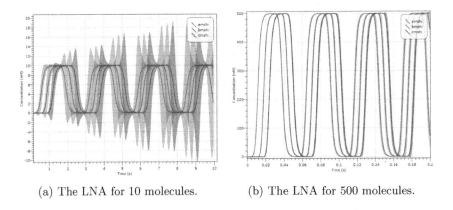

(a) The LNA for 10 molecules.   (b) The LNA for 500 molecules.

Fig. 6. LNA simulation of the queue pipeline. In these plots we show standard deviation, calculated through LNA, of an oscillatory pattern created by propagating a value of 1 and then 0. The maximum molecular count for each species in (a) is 10 while in (b) is 500. The variance decreases greatly with an increase in molecular count. We plot the values of *amshi, bmshi, cmshi*, which represent a value of 1. The troughs indicate when 0 is propagated.

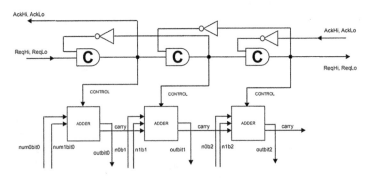

Fig. 7. Full Ripple Carry Adder used in conjunction with our Muller C-pipeline to stagger computation across the adders.

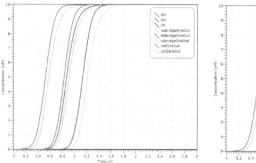 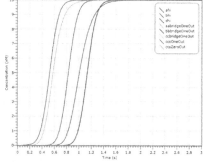

(a) Adder response to value of 101010.   (b) Adder response to value of 100010.

Fig. 8. Deterministic simulation of the adder circuit responding to various inputs. We overlay this with signals present in the pipeline used to coordinate the carry bit from each adder, represented by *aabridgeOneOut*, *bbbridgeOneOut* and *ccbridgeOneOut*. In (b), the final output signals cross due to pre-calculation by the adders before the carry bit arrives.

systems involving *Toehold Mediated Branch Migration and Strand Displacement* (DSD). DNA strand displacement has already been shown to be a universal substrate for chemical kinetics, specifically for bi-molecular reactions [26]. In addition to modelling the behaviours at the CRN level, we also implemented our CRN designs in two-domain DNA strand displacement devices [3] using the Visual DSD tool [16], thus providing further evidence of their experimental viability, at least for the construction of DNA-based devices.

6   Conclusion

We have proposed a novel design for an asynchronous computing device based on Chemical Reaction Networks. CRNs are inherently asynchronous, and thus particularly well suited to this computational paradigm. Our designs are based on a simple, bi-molecular reaction motif inspired by Approximate Majority [1,5], and assume well-mixed solution and constant, uniform rates. Moreover, they do not rely on the universal clock which is difficult to realise. Since an arbitrary CRN can be physically realised using DNA strand displacement [26], as recently demonstrated experimentally in [9], the proposed designs are in principle implementable, and we have confirmed this in theory by modelling them in the two-domain setting [3] using Visual DSD [16,22]. Our designs are the first feasible implementation of an asynchronous computing device in chemical kinetics and are relevant for a multitude of applications in nanotechnology and synthetic biology.

References

1. Angluin, D., Aspnes, J., Eisenstat, D.: A simple population protocol for fast robust approximate majority. Distrib. Comput. **21**(2), 87–102 (2008)
2. Bortolussi, L., Cardelli, L., Kwiatkowska, M., Laurenti, L.: Approximation of probabilistic reachability for chemical reaction networks using the linear noise approximation. In: Proceedings of 13th International Conference on Quantitative Evaluation of SysTems (QEST 2016). LNCS. Springer (2016) (to appear)
3. Cardelli, L.: Two-domain DNA strand displacement. Dev. Comput. Models **26**, 47–61 (2010)
4. Cardelli, L.: Morphisms of reaction networks that couple structure to function. BMC Syst. Biol. **8**(1), 84 (2014)
5. Cardelli, L., Csikász-Nagy, A.: The cell cycle switch computes approximate majority. Sci. Rep. **2**, 1–37 (2012)
6. Cardelli, L., Kwiatkowska, M., Laurenti, L.: Stochastic analysis of chemical reaction networks using linear noise approximation. In: Roux, O., Bourdon, J. (eds.) CMSB 2015. LNCS, vol. 9308, pp. 64–76. Springer, Heidelberg (2015)
7. Chen, H.-L., Doty, D., Soloveichik, D.: Deterministic function computation with chemical reaction networks. Nat. Comput. **13**(4), 517–534 (2013)
8. Chen, H.-L., Doty, D., Soloveichik, D.: Rate-independent computation in continuous chemical reaction networks. In: Proceedings of the 5th Conference on Innovations in Theoretical Computer Science, pp. 313–326. ACM (2014)
9. Chen, Y.-J., Dalchau, N., Srinivas, N., Phillips, A., Cardelli, L., Soloveichik, D., Seelig, G.: Programmable chemical controllers made from DNA. Nat. Nanotechnol. **8**(10), 755–762 (2013)
10. Cook, M., Soloveichik, D., Winfree, E., Bruck, J.: Programmability of chemical reaction networks. In: Condon, A., Harel, D., Kok, J.N., Salomaa, A., Winfree, E. (eds.) Algorithmic Bioprocesses, pp. 543–584. Springer, Heidelberg (2009)
11. Dannenberg, F., Kwiatkowska, M., Thachuk, C., Turberfield, A.J.: Dna walker circuits: computational potential, design and verification. Nat. Comput. **14**(2), 195–211 (2015)
12. de Silva, A.P., McClenaghan, N.D.: Molecular-scale logic gates. Chem.-A Eur. J. **10**(3), 574–586 (2004)
13. Gillespie, D.T.: Exact stochastic simulation of coupled chemical reactions. J. Phys. Chem. **81**(25), 2340–2361 (1977)
14. Hjelmfelt, A., Weinberger, E.D., Ross, J.: Chemical implementation of finite-state machines. Proc. Nat. Acad. Sci. **89**(1), 383–387 (1992)
15. Kwiatkowska, M., Norman, G., Parker, D.: PRISM 4.0: verification of probabilistic real-time systems. In: Gopalakrishnan, G., Qadeer, S. (eds.) CAV 2011. LNCS, vol. 6806, pp. 585–591. Springer, Heidelberg (2011)
16. Lakin, M.R., Youssef, S., Polo, F., Emmott, S., Phillips, A.: Visual DSD: a design and analysis tool for dna strand displacement systems. Bioinformatics **27**(22), 3211–3213 (2011)
17. Magnasco, M.O.: Chemical kinetics is Turing universal. Phys. Rev. Lett. **78**, 1190–1193 (1997)
18. Manohar, R., Martin, A.J.: Quasi-delay-insensitive circuits are Turing-complete. Technical report, DTIC Document (1995)
19. Napp, N.E., Adams, R.P.: Message passing inference with chemical reaction networks. In: Advances in Neural Information Processing Systems, pp. 2247–2255 (2013)

20. Nguyen, N.-P., Myers, C., Kuwahara, H., Winstead, C., Keener, J.: Design and analysis of a robust genetic Muller C-element. J. Theoret. Biol. **264**(2), 174–187 (2010)
21. Nguyen, N.-P.D., Kuwahara, H., Myers, C.J., Keener, J.P.: The design of a genetic Muller C-element. In: 13th IEEE International Symposium on Asynchronous Circuits and Systems, ASYNC 2007, pp. 95–104. IEEE (2007)
22. Phillips, A., Cardelli, L.: A programming language for composable DNA circuits. J. R. Soc. Interface **6**(Suppl 4), S419–S436 (2009)
23. Senum, P., Riedel, M.: Rate-independent constructs for chemical computation. PloS One **6**(6), e21414 (2011)
24. Shin, S.W.: Compiling and verifying DNA-based chemical reaction network implementations. Ph.D. thesis, California Institute of Technolog (2011)
25. Soloveichik, D., Cook, M., Winfree, E., Bruck, J.: Computation with finite stochastic chemical reaction networks. Nat. Comput. **7**(4), 615–633 (2008)
26. Soloveichik, D., Seelig, G., Winfree, E.: DNA as a universal substrate for chemical kinetics. Proc. Nat. Acad. Sci. **107**(12), 5393–5398 (2010)
27. Spars, J., Furber, S.: Principles Asynchronous Circuit Design. Springer, New York (2002)
28. Van Kampen, N.G.: Stochastic Processes in Physics and Chemistry, vol. 1. Elsevier, Amsterdam (1992)

Hierarchical Self-Assembly of Fractals with Signal-Passing Tiles

(Extended Abstract)

Jacob Hendricks[1], Meagan Olsen[2], Matthew J. Patitz[3(✉)], Trent A. Rogers[3], and Hadley Thomas[2]

[1] Department of Computer Science and Information Systems,
University of Wisconsin - River Falls, River Falls, WI, USA
`jacob.hendricks@uwrf.edu`
[2] Fayetteville High School, Fayetteville, AR, USA
`olsen.megs@gmail.com`, `hadleythomas88@gmail.com`
[3] Department of Computer Science and Computer Engineering,
University of Arkansas, Fayetteville, AR, USA
`{patitz,tar003}@uark.edu`

Abstract. In this extended abstract, we present high-level overviews of tile-based self-assembling systems capable of producing complex, infinite, aperiodic structures known as discrete self-similar fractals. Fractals have a variety of interesting mathematical and structural properties, and by utilizing the bottom-up growth paradigm of self-assembly to create them we not only learn important techniques for building such complex structures, we also gain insight into how similar structural complexity arises in natural self-assembling systems. Our results fundamentally leverage hierarchical assembly processes, and use as our building blocks square "tile" components which are capable of activating and deactivating their binding "glues" a constant number of times each, based only on local interactions. We provide the first constructions capable of building arbitrary discrete self-similar fractals at scale factor 1, and many at temperature 1 (i.e. "non-cooperatively"), including the Sierpinski triangle.

1 Introduction

Fractal patterns have mathematically interesting characteristics, such as recursive self-similarity, and structural properties which lend naturally occurring fractal structures, such as branch patterns and circulatory systems, impressive abilities to efficiently maximize coverage, dissipate heat, etc. Such fractal patterns in nature tend to arise via local processes following relatively simple sets

Matthew J. Patitz—This author's research was supported in part by National Science Foundation Grant CCF-1422152.

Trent A. Rogers—This author's research was supported by the National Science Foundation Graduate Research Fellowship Program under Grant No. DGE-1450079, and National Science Foundation Grant CCF-1422152.

Y. Rondelez and D. Woods (Eds.): DNA 2016, LNCS 9818, pp. 82–97, 2016.
DOI: 10.1007/978-3-319-43994-5_6

of rules, as forms of self-assembly. Because of this, and the complex aperiodic nature of fractals, they are a natural target of study during the development of artificial self-assembling systems. As one of the first mathematical abstractions of self-assembling systems, Winfree's abstract Tile Assembly Model (aTAM) [14] has been the platform for several results showing the impossibility of self-assembling discrete self-similar fractals such as the Sierpinski triangle[1][9] and similar fractals [1], and also for designing systems which can approximate them [9,10,13]. In a more generalized model called the 2-Handed Assembly Model [2,4] (2HAM, a.k.a. Hierarchical Assembly Model) which allows pairs of large assemblies to bind together, rather than being restricted to only single tile additions per step like the aTAM, the impossibility of self-assembling the Sierpinski triangle [2] has also been shown. In further generalizations allowing larger numbers of assemblies to combine in single steps [3] shapes were shown to self-assemble as well one unscaled fractal, the Sierpinski carpet.

A more recently developed model of tile-based self-assembly, called the Signal-passing Tile Assembly Model (STAM) was developed in [11] to model the behavior of DNA-based tiles capable of strand displacement reactions initiated during the binding of their glues which can then either activate or deactivate other glues on the same tile. Such signal-passing tiles have been experimentally demonstrated [12], and various theoretical results have demonstrated the power of systems using such tiles to efficiently simulate Turing machines [11], replicate patterns [8] and shapes [7], and also to self-assemble the Sierpinski triangle at scale factor 2.

In this extended abstract, we provide a high-level overview of constructions in the STAM which include: (1) the first capable of self-assembling the Sierpinski triangle at scale factor 1, which in fact even works at temperature 1 (i.e., a form of non-cooperative assembly), and (2) an algorithmic method which uses the definition of a fractal as input in order to develop an STAM system which self-assembles that fractal at scale factor 1. The second result develops systems at temperature 1 for an infinite class of fractals, and for the full class of discrete self-similar fractals at temperature 2. Our results fundamentally leverage techniques of hierarchical self-assembly and utilize specifically designed instances of geometric hindrance to allow fractals to grow in a carefully controlled, stage-by-stage manner. In the following sections, we first give an overview of the models and terminology used in the paper, then provide high-level, intuitive overviews of both main constructions. Full details of the constructions can be found in [6].

---

[1] In this paper we refer only to "strict" self-assembly, wherein a shape is made by placing tiles only within the domain of the shape, as opposed to "weak" self-assembly where a pattern representing the shape can be formed embedded within a framework of additional tiles.

2   Preliminaries

Here we provide informal descriptions of the models and terms used in this paper.

2.1   Informal Description of the STAM

The STAM, as formulated, is intended to provide a model based on experimentally plausible mechanisms for glue activation and deactivation. A detailed, technical definition of the STAM model is provided in [5].

(Note that the STAM is an extension of the 2HAM, and an informal description of the 2HAM can be found in [6]) In the STAM, tiles are allowed to have sets of glues on each edge (as opposed to only one glue per side as in the aTAM and 2HAM). Tiles have an initial state in which each glue is either "on" or "latent" (i.e. can be switched on later). Tiles also each implement a transition function which is executed upon the binding of any glue on any edge of that tile. The transition function specifies, for each glue g on a tile, a set of glues (along with the sides on which those glues are located) and an action, or *signal* which is *fired* by g's binding, for each glue in the set. The actions specified may be to: 1. turn the glue on (only valid if it is currently latent), or 2. turn the glue off (valid if it is currently on or latent). This means that glues can only be on once (although may remain so for an arbitrary amount of time or permanently), either by starting in that state or being switched on from latent (which we call *activation*), and if they are ever switched to off (called *deactivation*) then no further transitions are allowed for that glue. This essentially provides a single "use" of a glue (and the signal sent by its binding). Note that turning a glue off breaks any bond that glue may have formed with a neighboring tile. Also, since tile edges can have multiple active glues, when tile edges with multiple glues are adjacent, it is assumed that all matching glues in the on state bind (for a total binding strength equal to the sum of the strengths of the individually bound glues). The transition function defined for a tile type is allowed a unique set of output actions for the binding event of each glue along its edges, meaning that the binding of any particular glue on a tile's edge can initiate a set of actions to turn an arbitrary set of the glues on the sides of the same tile on or off.

As the STAM is an extension of the 2HAM, binding and breaking can occur between tiles contained in pairs of arbitrarily sized supertiles. It was designed to model physical mechanisms which implement the transition functions of tiles but are arbitrarily slower or faster than the average rates of (super)tile attachments and detachments. Therefore, rather than immediately enacting the outputs of transition functions, each output action is put into a set of "pending actions" which includes all actions which have not yet been enacted for that glue (since it is technically possible for more than one action to have been initiated, but not yet enacted, for a particular glue). Any event can be randomly selected from the set, regardless of the order of arrival in the set, and the ordering of either selecting some action from the set or the combination of two supertiles is also completely arbitrary. This provides fully asynchronous timing between the initiation, or firing, of signals (i.e. the execution of the transition function which puts them in the pending set) and their execution (i.e. the changing of the state of the target glue), as an arbitrary number of supertile binding events may occur before any signal is executed from the pending set, and vice versa.

An STAM system consists of a set of tiles and a temperature value. To define what is producible from such a system, we use a recursive definition of producible assemblies which starts with the initial tiles and then contains any supertiles which can be formed by doing the following to any producible assembly: 1. executing any entry from the pending actions of any one glue within a tile within that supertile (and then that action is removed from the pending set), 2. binding with another supertile if they are able to form a τ-stable supertile, or 3. breaking into 2 separate supertiles along a cut whose total strength is $< \tau$.

2.2 Discrete Self-Similar Fractals

We define $\mathbb{N}_g$ as the subset $\{0, 1, ..., g-1\}$ of $\mathbb{N}$, and if $A, B \subseteq \mathbb{N}^2$, then $A + (x, y)B = \{(x_a, y_a) + (x \cdot x_b, y \cdot y_b) | (x_a, y_a) \in A \text{ and } (x_b, y_b) \in B\}$. We then define discrete self-similar fractals as follows:

We say that $\mathbf{X} \subset \mathbb{N}^2$ is a *discrete self-similar fractal* (or *dssf* for short) if there exists a set $\{(0, 0)\} \subset G \subset \mathbb{N}^2$ where G is connected, $w_G = \max(\{x | (x, y) \in G\}) + 1$, $h_G = \max(\{y | (x, y) \in G\}) + 1$, w_G and $h_G > 1$, and $G \subsetneq \mathbb{N}_{w_G} \times \mathbb{N}_{h_G}$, such that $\mathbf{X} = \bigcup_{i=1}^{\infty} X_i$, where X_i, the ith stage of $\mathbf{X}$, is defined by $X_1 = G$ and $X_{i+1} = X_i + (w_G^i, h_G^i)G$. We say that G is the generator of $\mathbf{X}$. Essentially, the generator is a connected set of points in $\mathbb{N}^2$ containing $(0, 0)$, points at both $x > 0$ and $y > 0$, and is not a completely filled rectangle. Every stage after the constructor is composed of copies of the previous stage arranged in the same pattern as the generator.

A connected discrete self-similar fractal is one in which every component is connected in every stage, i.e. there is only one connected component in the grid graph formed by the points of the shape.

Figure 1 shows, as an example, the first 4 stages of the discrete self-similar fractal known as the Sierpinski triangle. In this example, $G = \{(0, 0), (1, 0), (0, 1)\}$.

We also define a subset of connected discrete self-similar fractals which we call *singly-concave* as containing any connected discrete self-similar fractal $\mathcal{F}$ such that, if stage 2 of $\mathcal{F}$, $\mathcal{F}_2$, is contained within a bounding box, on the straight line path p from any point on the bounding box into the first location adjacent to $\mathcal{F}_2$,

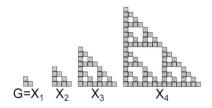

$G=X_1$ X_2 X_3 X_4

Fig. 1. Example discrete self-similar fractal: the first 4 stages of the Sierpinski triangle

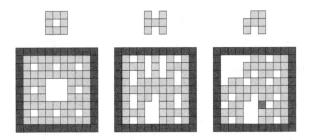

Fig. 2. Three example generators, and their associated second stages contained within bounding boxes (darker grey). The left two are singly-concave because any paths (examples in yellow) from the bounding boxes to the edge of their second stage tiles meet only contiguous sets of edges of the fractal. The rightmost isn't because the red and yellow tiles of one path meet two non-contiguous sets. (Color figure online)

the set of all edges along which p is adjacent to $\mathcal{F}_2$ are contiguous. Intuitively, singly-concave fractals do not have concavities which occur within the "sides" of other concavities. Examples can be seen in Fig. 2.

3   Strict Self-Assembly of the Discrete Sierpinski Triangle

Theorem 1. *There exists an STAM system $\mathcal{T}_\Delta = (T, 1)$ such that $\mathcal{T}_\Delta$ has exactly one infinite terminal supertile α_Δ, and* dom $(\alpha_\Delta) = S_\Delta$, *i.e. is exactly the discrete Sierpinski triangle, and for all $\alpha \in \mathcal{A}_\square[T]$ such that $\alpha \neq \alpha_\Delta, |\text{dom}(\alpha)| \leq 4$.*

Proof. We prove Theorem 1 by construction, and thus present an STAM tile assembly system $\mathcal{T}_\Delta$ and show that it strictly self assembles S_Δ, while any assemblies which detach from the assembly (or otherwise form) during its growth (which we call "junk" assemblies) all become terminal at sizes ≤ 4. At a high level, $\mathcal{T}_\Delta$ uses 2HAM principles (i.e. combinations of large supertiles) to combine a northern, southern, and western version of each stage n for $1 < n < \infty$ through geometric matching, to produce stage $n + 1$. Here, we provide a brief overview. Details of the construction can be found in [6]

From a "hard-coded" start at stage two (i.e. base tiles initially combine to form this stage before allowing formation of subassemblies), each stage n must completely grow before the subassemblies that make that stage are able to combine and form the next stage, $n + 1$. Only when an assembly representing stage n is completely built can an *initiator* tile attach to it and turn on a specific glue that allows for nondeterministic binding of one of three tiles, which then tells that copy of stage n to become one of three substages for stage $n + 1$. By definition of the Sierpinski triangle, there are three substages of each stage that correspond to the three points in the generator, i.e. $(0, 0)$, $(1, 0)$, and $(0, 1)$. The nondeterministic binding of one of the initiator types initiates an assembly sequence which grows either a *tooth* or *gap* on the assembly to which it is attached.

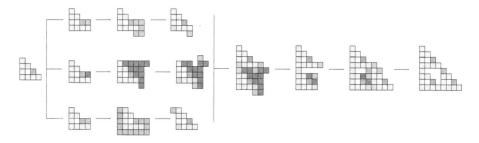

Fig. 3. High-level formation process of the stage of S_Δ immediately following the initial stage two formation. From top to bottom are shown $S_{\Delta s}$, $S_{\Delta u}$, and $S_{\Delta w}$. (Color figure online)

A tooth is a one-tile protrusion from a flat surface, and a gap is a one-tile cavity in a flat surface (see Fig. 3 for examples). One produces a southern tooth and becomes the northwest portion of S_Δ stage $n + 1$, a second produces a western tooth and becomes the southeast portion of S_Δ stage $n + 1$, and the third goes through two main phases, first filling in along the diagonal to make roughly a square with a gap in the north face, then after connecting to the northern piece opens a gap in the east to allow its connection to the eastern piece. We will call the substage assemblies $S_{\Delta s}$, $S_{\Delta w}$, and $S_{\Delta u}$, respectively, and the tile sets (which are subsets of T) that form them $T_{\Delta s}$, $T_{\Delta w}$, and $T_{\Delta u}$. Note that the glues that allow connections of the substages are only activated after the necessary geometry is in place to verify the sizes of the complementary pieces, and after the substage connections, all tiles not within the domain of stage n fall off of the assembly.

As depicted in Fig. 3, $S_{\Delta s}$ and $S_{\Delta u}$ are the first subassemblies to combine with attachment points on the southwest and northwest corners of their respective assemblies. The southern tooth of $S_{\Delta s}$, depicted in green, aligns with the slot created in the $S_{\Delta u}$ assembly, shown in purple. The two assemblies can only align when they are of the proper size due to the orange blocker tile located to the immediate right of the $S_{\Delta u}$ slot. $S_{\Delta u}$ cannot decay appropriately (i.e. cause "unwanted" tiles to fall off) until this blocker tile is in place; only after the blocker tile binds does a series of glues activate that result in the removal of the gap tile. After a sequence of detachments removes the uppermost row and eastmost column of the resulting assembly, a second blocker tile attaches to the southeastern corner of the $S_{\Delta s}$ assembly, finishing the decay and enabling the alignment of the $S_{\Delta w}$ assembly of the same stage. A final series of decay removes all other tiles that do not fit with the formation of stage n of the Sierpinski triangle.

Assembly of substages and complete stages after the first combination depicted in Fig. 4 follow the same general pattern of creation and decay, with few notable differences. The formation of the S_Δ subassemblies does require slightly more intricate systems of signals, largely to create the stair-step mechanism seen in $S_{\Delta s}$ and $S_{\Delta w}$.

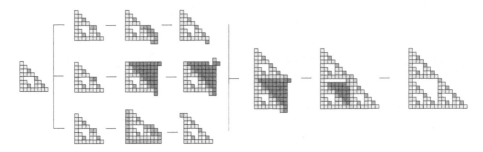

Fig. 4. High-level formation process of the fourth stage of S following formation of stage three. From top to bottom are shown $S_{\Delta s}$, $S_{\Delta u}$, and $S_{\Delta w}$.

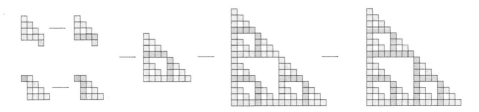

Fig. 5. Base tiles that carry either vertical or horizontal signals are depicted in gray; the location of the signals demonstrate that no tiles used to carry either vertical or horizontal signals are used in the same manner again, i.e. to pass signals during the formation of more than one stage.

Due to STAM properties that maintain that no tile can send its signals more than once, care has been taken to ensure that no signals sent through the tiles of the $\mathcal{T}_{\Delta}$ tile set are used more than once. As shown in Fig. 5, the two regions that send signals through the base tiles are never in a location to be used for the same horizontal or vertical signal paths more than once.

Throughout the assembly process, junk tiles and subassemblies are continuously removed with the help of *blocker* tiles. Figure 6 displays this process for $S_{\Delta w}$. Each junk assembly removes itself only when the appropriate signals have been passed through it and, for many assemblies, a corresponding blocker tile has attached. This prevents active bonds that cannot be guaranteed deactivation

Fig. 6. Blocker tiles depicted in orange function alongside the blue tiles that make up the $T_{\Delta w}$ set to ensure that all potential junk tiles do not negatively affect the assembly.

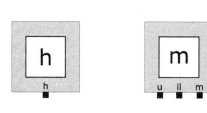
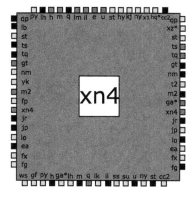

Fig. 7. Comparison of two helper tiles, h and m, alongside $T_{\Delta u}$ tile $xn4$, after it has attached to a subassembly and then subsequently detached, displays similar glues that are on. All glues that are on for the $xn4$ tile have either had their signals used, as is the case for the m, il, h, and u glues, or are not capable of interacting with subassemblies due to their counterparts turning on in isolation.

within the asynchronous STAM model from potentially interfering with active constructions. By binding a blocker tile, junk assemblies are created with no volatile perimeter glues. Blocker tiles also function to change the geometry of the junk assemblies to prevent any interference.

It is worth noting that some junk assemblies, particularly within the $T_{\Delta u}$ set, are capable of interacting with T_{Δ} subassemblies at various stages. The glues that are capable of interaction on these assemblies, however, have already had their signals used and function like existing blocker and helper tiles. This means that their interaction does not result in a negative impact on the assembly as a whole, instead assisting in the proper formation of S_{Δ} stages.

Figure 7 depicts an example of the previously described scenario. In this case, the $xn4$ tile functions like either the m or h tile, depending on which of its glues bind. The only glues that remain exposed in junk assemblies are either capable of performing a similar function, or do not interact with subassembly formation due to their counterpart glues turning on in isolation (i.e. a western counterpart for an eastern glue turns on only when a tile has attached to the eastern face of the tile in question).

In this process in which three separate versions of any assembly at stage n form and combine through the creation and alignment of teeth and gaps to ensure proper size integration to produce stage $n + 1$, and as throughout the process of assembly the junk assemblies are detaching in constant sized pieces that will not interact with the assembly in a negative manner, the correct strict self assembly of the Sierpinski triangle at scale 1 is produced.

4 Self-Assembly of Arbitrary Discrete Self-Similar Fractals: An Overview

We now state and provide a proof sketch of the two main generalized results of this paper. The proofs of the two results are similar enough that we provide a single proof sketch for both theorems.

Theorem 2. *For any connected discrete self-similar fractal $\mathcal{F}$, there exists an STAM system $\mathcal{T}_\mathcal{F} = (T, 2)$ such that $\mathcal{T}_\mathcal{F}$ has exactly one infinite terminal supertile α, and $\mathrm{dom}\,(\alpha) = \mathcal{F}$, i.e. is exactly the discrete self-similar fractal $\mathcal{F}$, and for all $\gamma \in \mathcal{A}_\square[T]$ such that $\gamma \neq \alpha, |\mathrm{dom}\,(\gamma)| \leq 2$.*

Theorem 3. *For any connected discrete self-similar fractal $\mathcal{F}$ which is singly-concave, there exists an STAM system $\mathcal{T}_\mathcal{F} = (T, 1)$ such that $\mathcal{T}_\mathcal{F}$ has exactly one infinite terminal supertile α, and $\mathrm{dom}\,(\alpha) = \mathcal{F}$, i.e. is exactly the discrete self-similar fractal $\mathcal{F}$, and for all $\gamma \in \mathcal{A}_\square[T]$ such that $\gamma \neq \alpha, |\mathrm{dom}\,(\gamma)| \leq 2$.*

Proof (Proof Sketch). The proofs of Theorems 2 and 3 are very nearly identical, with slight modifications in only a few places, so we now give an overview of the self-assembly of arbitrary discrete self-similar fractals. A detailed version of the construction can be found in [6]. We describe the self-assembly of an arbitrary fractal $\mathcal{F}$ generated by a generator G. The general idea of the construction is that we grow $\mathcal{F}$ in a stage-by-stage manner, making sure that the assembly of each stage is complete before it is able to join in the formation of the next stage. However, the growth of an infinite number of all stages is happening in parallel, so we must be careful to ensure that no copies of differing stages can bind or interfere with each other.

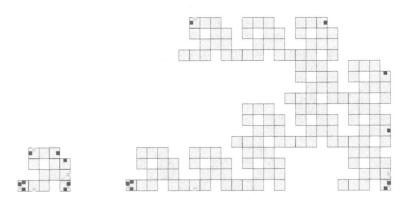

Fig. 8. An example of a generator and stage 2 with the locations of **end** (blue), **preconnect** (yellow), **init** (aqua), and **preinit** (purple) glues shown. Note that those positions are first marked in stage 2 and not in the generator, but they are shown here with markings to demonstrate their locations relative to the generator. (Color figure online)

Our construction begins by "hard-coding" tiles for stage 2 of $\mathcal{F}$. We mark the perimeter with special *marker* glues in locations which: (1) are the extreme points of each side, a total of 8 of these; (2) are the locations in each of N,E,S, and W which are used to form connections between copies of the same stage i which are combining to form stage $i+1$, there are 4 of these; and (3) a location where a special *initiator* tile type can attach to an assembly of a completed stage to allow it to begin the process of growth into a substage of the next stage. We call the first type of glues `end` glues, the second type of glues `preconnect` glues which are replaced by `connect` glues, and the third type of glues `preinit` glues which are later replaced by `init` glues. An example of the locations of these glues is shown in Fig. 8. The recursive structure of fractals allows us to treat each stage i as a combination of copies of stage $i-1$, where each copy logically represents one of the points of the generator. Our construction ensures that each completed copy of a stage has an identical set of special marker glues on its perimeter (interspersed by increasing numbers of generic edge marking glues), and is thus able to perform the same algorithmic steps on each to grow subsequent stages.

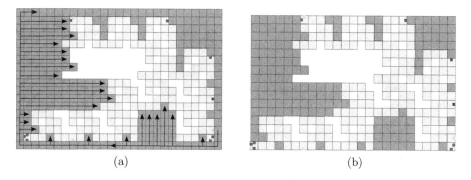

(a)   (b)

Fig. 9. Example of how sides of a substage are filled in by a perimeter path. The blue rectangles represent the location of `end` glues, the yellow rectangles represent the location of `connect` glues, and the aqua rectangles represent the `init` glue. (Color figure online)

An assembly which has completed formation of stage i undergoes a process of *differentiation* when one of the initiator tiles attaches (of which there are $|G|$ types, one for each point in the generator), during which its marker glues are transformed to match those necessary for it to become the substage location in stage $i+1$ corresponding to its initiator type. This also allows it to combine, one at a time, to the other necessary substage assemblies, in a well-defined ordering. The combinations of stages are controlled by not only the special connection glues and tooth-and-gap geometries similar to those used in the Sierpinski triangle construction, but also enabled by the formation of perimeter paths and filler growth (as shown in Fig. 9) which are important in creating the necessary geometry (i.e. smooth surfaces). The perimeter path and filler tiles grow around the substage assembly and activate any **connect** glues needed for connecting to other substage assemblies. They do this in such a way that the **connect** glue is not activated until the filler tiles are present (in order to make the side "smooth") and the tooth tile is present (which prevents incorrect substage assemblies

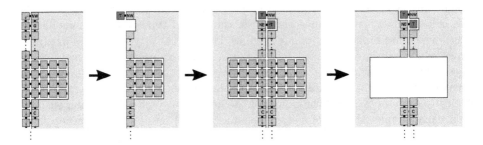

Fig. 10. An example of the growth process of a tooth and gap on the western side of a supertile preparing to bind to another supertile along that side.

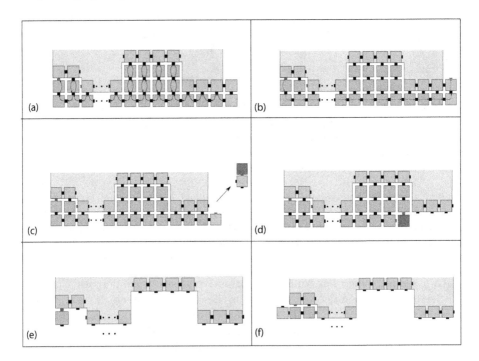

Fig. 11. A schematic which shows an overview of the process by which the perimeter path and filler tiles dissociate for the case where the side of the stage on which the perimeter path tiles are growing does not need to form a connection between another stage. The blue glues represent the **end** glues. The red glues represent the deactivator glues. The lines and arrows show how signals propagate. (Color figure online)

from binding). An example of this is shown in Fig. 10. In the case that $\mathcal{F}$ is singly-concave, the system we are designing has $\tau = 1$ and the filler tiles probe for the presence of the side of the substage assembly with τ strength glues. Note that by the assumption that $\mathcal{F}$ is singly-concave, we are guaranteed that there is a preceding row of tiles which prevents the erroneous binding of a "smaller" substage to the exposed glue since such a smaller substage assembly would have to be offset into that proceeding row by at least one position to allow the necessary glues to align (since the glues on the corners are different than those not

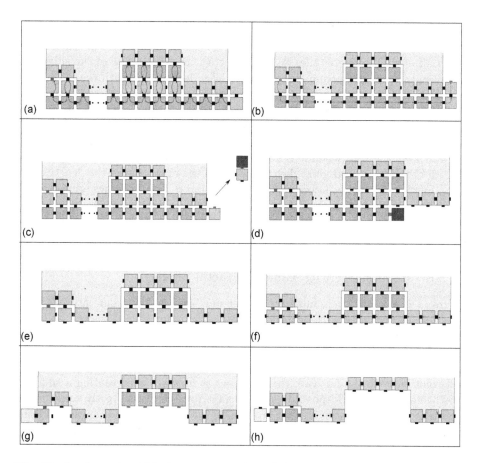

Fig. 12. A schematic which shows an overview of the process by which the perimeter path and filler tiles dissociate for the case where the side of the substage assembly on which the perimeter path tiles are growing does need to form a connection between another substage assembly. The blue glues represent the **end** glues. The red glues represent the deactivator glues. The lines and arrows show how signals propagate. (Color figure online)

on the corners). In the case that $\mathcal{F}$ is not singly-concave, then the system is at temperature 2 and the filler tiles probe for the presence of the side of the substage assembly with a strength 1 glue which means that "smaller" substage assemblies cannot bind due to insufficient binding strength.

Figure 9 shows how the perimeter path and filler tiles start at the init glue and travel around the assembly in a counterclockwise direction. After the perimeter path and filler tiles assemble, the tiles need to dissociate. During the growth of these tiles, *deactivator* glues activate in such a way that they are "hidden" by the tiles to which they bind. Whenever the perimeter path and filler tiles receive a signal to dissociate, a signal is propagated through the perimeter path tiles which causes a deactivator glue to be exposed on the tile which started the growth of the perimeter path tiles on the side. This causes a chain reaction in which all of the perimeter tiles decay one at a time. If the side of the substage assembly is not a connection side, the filler tiles decay along with the perimeter path tiles. Otherwise, the perimeter path tiles decay, another substage assembly binds to the side, and then the filler tiles decay via a signal that begins propagating at the connect glue. Figure 11 shows an example of the perimeter path and filler tiles decaying in the case that the side is not a connection side (while Fig. 12 shows the case when it is a connection side). These tiles are designed so that all of their exposed glues deactivate (with the exception of the deactivator glue). Furthermore, the majority of these tiles bind to "static" glues (glues which do not trigger events) on the substage assembly. This means that if a tile that is in the process of deactivating binds to the substage assembly, it will not trigger any unwanted binding events.

4.1 Errors at $\tau = 1$ with General Fractal Shapes

Figure 13 shows why there must be a difference between the temperature-2 construction which works with general discrete self-similar fractal shapes, and the $\tau = 1$ construction which only works for a constrained subset of fractal shapes. Specifically, if a row of filler tiles can grow to a position where there is nothing below it, but its tiles also have active glues which are able to detect their eventual collision with the far wall, then at $\tau = 1$ a supertile representing a smaller substage assembly could possibly bind to that row, thus incorrectly connecting supertiles of different stages. This problem is avoided in the $\tau = 2$ construction by making these glues strength-1, thus preventing them from having enough strength to bind the supertiles, along with the fact that only one such row can be growing at a time and thus exposing such glues, and the pattern of the eventual decay of those filler rows is also carefully designed to prevent the ability for erroneous binding.

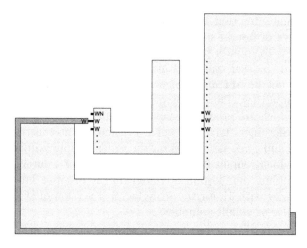

Fig. 13. A schematic depiction of the error that could occur in a $\tau = 1$ system in which the fractal shape contains certain types of concavities.

5 Conclusion

In this paper, we have shown how signal-passing tiles can be designed to self-assemble discrete self-similar fractals in a very hierarchical manner. That is, separately and in parallel, the copies of each stage of a given fractal self-assemble, and then those copies combine to form the next stage. During this process, the sides of each copy which must combine with each other are prepared for that combination in such a way that only copies of the same stage can combine, and only once all pieces of the previous stage have attached. This is done by using geometric hindrance created by small bumps and dents at carefully spaced locations, along with well-controlled timing of glue activations. Once copies of a stage have bound together, any "filler" tiles which attached to create the necessary geometry along the combining edge, but which are not part of the final fractal shape, then detach. These detached tiles, called "junk", are designed so that they always eventually become subassemblies which are no greater than size 4, and they cannot interfere with any additional assembly.

While these results allow for the self-assembly of fractals without any increase in scale factor and, for an infinite set of fractals, at temperature 1, there remain a number of improvements which may be possible and open questions which remain. First, is the cooperativity of temperature 2 strictly necessary for fractals which are not singly concave? I.e., is there a temperature 1 construction which can correctly self-assemble such fractals? It appears that such a construction would require non-trivial adaptations to or departures from our current construction. Conversely, proving it impossible also appears to be difficult. Second, can the maximum size of the terminal junk assemblies be reduced even further, to 1, 2, or 3? This also appears difficult due to the fact that within the STAM glue deactivations (like activations) happen asynchronously, and therefore it is not possible to guarantee that both sides of a bound glue (i.e. the glues on the two adjacent edges of bound tiles) have been deactivated before a tile or sub-assembly detaches. To counter problems that may arise from this, we frequently ensure that special "blocker" tiles first attach to the soon-to-be-junk tiles to hide glues which may not be `off` and which may allow the junk assembly to interfere with other supertiles. Third, is it possible to further reduce the number of junk assemblies which are produced, especially during the self-assembly of fractals such as the Sierpinski triangle, which in the current construction requires a number of tiles and junk assemblies on the order of approximately 1/4 of the number of tiles which remain to form each given stage? Fourth, our construction for self-assembling the Sierpinski triangle in the proof of Theorem 1 uses 48 tile types (compared with 19 for the scale factor 2 version of [11]). Can this be reduced to a similar or smaller number?

References

1. Barth, K., Furcy, D., Summers, S.M., Totzke, P.: Scaled tree fractals do not strictly self-assemble. In: Ibarra, O.H., Kari, L., Kopecki, S. (eds.) UCNC 2014. LNCS, vol. 8553, pp. 27–39. Springer, Heidelberg (2014)
2. Cannon, S., Demaine, E.D., Demaine, M.L., Eisenstat, S., Patitz, M.J., Schweller, R.T., Summers, S.M., Winslow, A.: Two hands are better than one (up to constant factors): self-assembly in the 2ham vs. atam. In: Portier, N., Wilke, T. (eds.) STACS. LIPIcs, vol. 20, pp. 172–184. Schloss Dagstuhl - Leibniz-Zentrum fuer Informatik (2013)
3. Chalk, C.T., Fernandez, D.A., Huerta, A., Maldonado, M.A., Schweller, R.T., Sweet, L.: Strict self-assembly of fractals using multiple hands. Algorithmica, 1–30 (2015)
4. Cheng, Q., Aggarwal, G., Goldwasser, M.H., Kao, M.-Y., Schweller, R.T., de Espanés, P.M.: Complexities for generalized models of self-assembly. SIAM J. Comput. 34, 1493–1515 (2005)
5. Fochtman, T., Hendricks, J., Padilla, J.E., Patitz, M.J., Rogers, T.A.: Signal transmission across tile assemblies: 3d static tiles simulate active self-assembly by 2d signal-passing tiles. Nat. Comput. 14(2), 251–264 (2015)
6. Hendricks, J., Olsen, M., Patitz, M.J., Rogers, T.A., Thomas, H.: Hierarchical self-assembly of fractals with signal-passing tiles (extended abstract). Technical Report 1606.01856, Computing Research Repository (2016)
7. Hendricks, J., Patitz, M.J., Rogers, T.A.: Replication of arbitrary hole-free shapes via self-assembly with signal-passing tiles. In: Calude, C.S., Dinneen, M.J. (eds.) UCNC 2015. LNCS, vol. 9252, pp. 202–214. Springer, Heidelberg (2015)
8. Keenan, A., Schweller, R., Zhong, X.: Exponential replication of patterns in the signal tile assembly model. Nat. Comput. 14(2), 265–278 (2015)
9. Lathrop, J.I., Lutz, J.H., Summers, S.M.: Strict self-assembly of discrete Sierpinski triangles. Theoret. Comput. Sci. 410, 384–405 (2009)
10. Lutz, J.H., Shutters, B.: Approximate self-assembly of the sierpinski triangle. Theory Comput. Syst. 51(3), 372–400 (2012)
11. Padilla, J.E., Patitz, M.J., Schweller, R.T., Seeman, N.C., Summers, S.M., Zhong, X.: Asynchronous signal passing for tile self-assembly: fuel efficient computation and efficient assembly of shapes. Int. J. Found. Comput. Sci. 25(4), 459–488 (2014)

12. Padilla, J.E., Sha, R., Kristiansen, M., Chen, J., Jonoska, N., Seeman, N.C.: A signal-passing dna-strand-exchange mechanism for active self-assembly of dna nanostructures. Angewandte Chemie Int. Ed. **54**(20), 5939–5942 (2015)
13. Patitz, M.J., Summers, S.M.: Self-assembly of discrete self-similar fractals. Nat. Comput. **1**, 135–172 (2010)
14. Winfree, E.: Algorithmic self-assembly of DNA. Ph.D. thesis, California Institute of Technology, June 1998

Resiliency to Multiple Nucleation in Temperature-1 Self-Assembly

Matthew J. Patitz[1]([✉]), Trent A. Rogers[1], Robert T. Schweller[2],
Scott M. Summers[3], and Andrew Winslow[4]

[1] Department of Computer Science and Computer Engineering,
University of Arkansas, Fayetteville, USA
{patitz,tar003}@uark.edu

[2] Department of Computer Science, University of Texas–Rio Grande Valley,
Edinburg, USA
robert.schweller@utrgv.edu

[3] Computer Science Department, University of Wisconsin–Oshkosh, Oshkosh, USA
summerss@uwosh.edu

[4] Département d'Informatique, Université libre de Bruxelles, Brussels, Belgium
awinslow@ulb.ac.be

Abstract. We consider problems in variations of the two-handed abstract Tile Assembly Model (2HAM), a generalization of Erik Winfree's abstract Tile Assembly Model (aTAM). In the latter, tiles attach one-at-a-time to a *seed*-containing assembly. In the former, tiles aggregate into *supertiles* that then further combine to form larger supertiles; hence, constructions must be robust to the choice of seed (nucleation) tiles. We obtain three distinct results in two 2HAM variants whose aTAM siblings are well-studied.

In the first variant, called the *restricted glue 2HAM* (rg2HAM), glue strengths are restricted to -1, 0, or 1. We prove this model is Turing universal, overcoming undesired growth by breaking apart undesired computation assembly via repulsive forces.

In the second 2HAM variant, the *3D 2HAM* (3D2HAM), tiles are (three-dimensional) cubes. We prove that assembling a (roughly two-layer) $n \times n$ square in this model is possible with $O(\log^2 n)$ tile types. The construction uses "cyclic, colliding" binary counters, and assembles the shape non-deterministically. Finally, we prove that there exist 3D2HAM systems that only assemble infinite *aperiodic* shapes.

M.J. Patitz—This author's research was supported in part by National Science Foundation Grant CCF-1422152.

T.A. Rogers—This author's research was supported by the National Science Foundation Graduate Research Fellowship Program under Grant No. DGE-1450079, and National Science Foundation Grant CCF-1422152.

R.T. Schweller—This author's research was supported in part by National Science Foundation Grants CCF-1117672 and CCF-1555626.

© Springer International Publishing Switzerland 2016
Y. Rondelez and D. Woods (Eds.): DNA 2016, LNCS 9818, pp. 98–113, 2016.
DOI: 10.1007/978-3-319-43994-5_7

1 Introduction

Self-assembly is the process through which a group of discrete components combine according to simple and local interaction rules to form a complex final structure. Taking inspiration from the many examples of self-assembly exhibited in nature, researchers are now investigating the use of nanoscale self-assembly for systematic nano-fabrication of atomically-precise computational, biomedical, and mechanical devices.

In the early 1980s, Ned Seeman [21] developed an experimental technique for such fabrication known as *DNA tile self-assembly*. In DNA tile self-assembly, a small number of single strands of DNA are used to form a DNA *tile* with four "sticky ends" consisting of short sequences of unpaired nucleotides, one for each of the cardinal directions: north, east, south, and west. A sticky end of one tile binds with a sticky end of a second tile if their nucleotides are Watson-Crick complements; more generally, multi-tile *supertiles* bind together similarly. Careful design of sticky ends enables an experimenter to program sets of DNA tiles to self-assemble into target structures.

Erik Winfree's abstract Tile Assembly Model (aTAM) is a discrete mathematical model of DNA tile self-assembly [24]. The aTAM abstracts DNA tiles are as translatable, but not rotatable, square tiles whose sides have alpha-numerically labeled *glues* with integer *strengths*. Two tiles or assemblies placed adjacently *bind* if the sums of the strengths of matching glues on coincident sides is at least a specified minimum threshold, called the *temperature* of the system. Self-assembly in the aTAM starts from a unique *seed* tile type and proceeds nondeterministically and asynchronously by single-tile addition to the growing *seed assembly*.

Two-Handed Tile Assembly. A well-studied seedless generalization of the aTAM is the *two-handed abstract Tile Assembly Model (2HAM)*. In the 2HAM, growth does not begin at a unique seed tile type. Instead, all possible pairs of tiles bind, followed by all possible pairs of *supertiles*, until no pair of resulting supertiles can bind further.

The role of temperature in both aTAM and 2HAM systems is critical, as it determines the criteria by which supertiles bind: a higher temperature defines a stronger binding criterion. At temperature 2, *cooperative* binding can be used to synchronize assembly and is known to confer complex algorithmic behavior in both the aTAM [1,17,23,24] and 2HAM [4]. On the other hand, the computational and geometric power of the temperature-1 aTAM (and 2HAM) famously remains open [10,14].

The difficulty of implementing cooperative binding in experimental DNA tile self-assembly has motivated the study of variants of the temperature-1 aTAM augmented with more practical features that confer similar capabilities. This work has established the temperature-1 aTAM is capable of universal computation and efficient shape assembly if the model is augmented with negative-strength glues [16], a third spatial dimension [7], "triggered" glues [15], or non-square tile shapes [11].

Unfortunately, these variations suffer from a common practical concern: avoiding "spurious nucleation", i.e. binding away from the seed tile. Indeed, even preventing spurious nucleation *with* cooperative binding has substantial challenges [2,6,18–20]. The difficulty of implementing aTAM-like seeded growth in experimental systems implies that, at least currently, the behavior of experimental DNA tile self-assembly systems without cooperative binding is captured better by the (temperature-1) 2HAM than the aTAM. Due in part to the newness of the 2HAM, prior study of augmented variants of the temperature-1 2HAM is limited to the staged model [8], a powerful "multi-pot" model.

Our Results. Here we obtain three positive results on the computational and geometric behaviors possible in two variants of the temperature-1 2HAM. Since the 2HAM permits growth to begin with any pair of tiles, positive results are necessarily robust for "multiple nucleation" errors.

In the first variant we consider, called the *restricted glue 2HAM* or *rg2HAM*, glue strengths are restricted to -1, 0, or 1. This is the two-handed equivalent of the restricted glue aTAM (rgTAM) [16]. We prove the rg2HAM is Turing universal (Sect. 3), demonstrating that the "anticooperative" behavior of the rg2HAM, like the cooperative behavior of the aTAM and 2HAM, is capable of simulating any Turing machine computation. The construction critically uses negative-strength glues to "break apart" nucleations encoding incorrect machine computations.

Note that the technique for proving the Turing universality of the temperature-1 3D aTAM [7] cannot be used in the 2HAM, since it uses a long path of tiles that branches and has "incorrect" branches blocked by a previous portion of the path. In the (seedless) 2HAM, an incorrect branch may assemble and become a "junk" assembly. In fact, every temperature-1 3D aTAM construction in [7] has similar problems.

In the second 2HAM variant, the *3D 2HAM* or *3D2HAM*, tiles are (three-dimensional) cubes. This is the two-handed equivalent of the 3D aTAM [7]. We prove that at temperature 1, the 3D2HAM is capable of efficient assembly of $n \times n$ squares (Sect. 4) and assembly of infinite aperiodic shapes (Sect. 5), matching known results in the (temperature-1 3D) aTAM [7].

The efficient square construction uses two layers in the third dimension and $O(\log^2 n)$ tile types. The key idea is a special "cyclic" binary counter that prevents incorrect growth from sabotaging completion of the counter's growth.

The aperiodic construction yields only infinite terminal supertiles whose shapes are not translations of themselves, i.e. have no repeating or periodic structure. Prior negative results on assembling aperiodic structures in the temperature-1 aTAM was given as evidence against the Turing universality of that model [10]. Moreover, the two-dimensional temperature-1 2HAM's ability to nucleate growth at any tile in an assembly has recently been shown to imply strong "pumping" results [5,9] that imply aperiodic systems do not exist.

Here, we contrast these results with the construction of a 3D2HAM system that assembles only infinite aperiodic assemblies. The construction simulates a scaled-up version of a standard aTAM binary counter and special "vacuum" glues to attach infinite periodic rows of this counter to completed (aperiodic) counters, yielding only aperiodic terminal supertiles.

The definition of "aperiodic" used and landscape of results related to aperiodic tile self-assembly systems closely match those of plane tilings: non-overlapping coverings of the plane using collections of shapes called *prototiles* (see [13]). A plane tiling is *aperiodic* provided it has no translational symmetry and a tile set is *aperiodic* provided every plane tiling it admits is aperiodic. Determining whether a prototile set admits a tiling is undecidable; as a corollary, some prototile sets admit only aperiodic tilings [3] (matching [24]). A long-standing conjecture states there are no aperiodic prototile sets from a restricted class of prototile sets, namely singleton tile sets, [12] (matching [17]). This conjecture was recently proved to not be true if a third dimension is allowed [22] (matching [7]).

Turing universality in both tile self-assembly and plane tiling implies the existence of aperiodic instances, but no implications in the other direction are known. Regardless, aperiodic behavior is generally considered evidence for Turing universality. Moreover, such behavior constrains possible proofs of Turing non-universality to those that do not forbid infinite non-repeating behavior.

2   Definitions

The set of *unit vectors* is $U_2 = \{(0,1),(1,0),(0,-1),(-1,0)\}$, also referred to as N, E, S, W, respectively. A *grid graph* is an undirected graph $G = (V,E)$ in which $V \subseteq \mathbb{Z}^2$ and every edge $\{a,b\} \in E$ has the property that $a - b \in U_2$.

Intuitively, a tile type t is a unit square that can be translated, but not rotated, and has a well-defined "side u" for each $u \in U_2$. Each side u of t has a *glue* with *label* $label_t(u)$ from some fixed alphabet and a non-negative integer *strength* $str_t(u)$ determined by its type. Two tiles t and t' that are placed at the points a and $a + u$ respectively, *bind* with *strength* $str_t(u)$ if $(label_t(u), str_t(u)) = (label_{t'}(-u), str_{t'}(-u))$ and with strength 0 otherwise.

2.1   Two-Handed Tile Assembly Model

The 2HAM is a generalization of the aTAM where any pair of multi-tile assemblies with sufficient binding strength can attach to each other. Included here is an informal description of the 2D 2HAM; see [4] for a more complete set of definitions.

A *supertile* is the equivalence class of all translations of an assembly, i.e. a "position-less" assembly.[1] The *binding graph* of a supertile is a weighted grid graph whose vertices are tiles and edges between adjacent tiles have weights corresponding to the strength of the binding between them. A supertile is τ-*stable* provided every cut of its binding graph has strength at least τ.

A 2HAM *tile assembly system* (TAS) is a pair $\mathcal{T} = (T,\tau)$, where T is a finite tile set and τ is the *temperature*; typically $\tau \in \{1,2\}$. Given a TAS $\mathcal{T} = (T,\tau)$,

---

[1] Such a distinction is only needed in two-handed models, where the seed cannot be used as a "reference point".

a supertile α is *producible*, denoted $\alpha \in \mathcal{A}[\mathcal{T}]$, provided that either α is a single tile in $\mathcal{T}$, or α is the union of two smaller non-overlapping producible supertiles α_1 and α_2 (called *subassemblies*) such that the cut of α into α_1 and α_2 has strength at least τ. For brevity, this relationship between α_1, α_2, and α is (non-uniquely) denoted $\alpha = \alpha_1 + \alpha_2$. A producible supertile α is *terminal*, denoted $\alpha \in \mathcal{A}_\square[\mathcal{T}]$, provided α cannot attach τ-stably to any other producible supertile.

A TAS is *directed* provided it has a unique terminal supertile. Given a connected shape $X \subseteq \mathbb{Z}^2$, we say a TAS $\mathcal{T}$ *self-assembles* X if the shape of every terminal supertile of $\mathcal{T}$ is a translation of X.

2.2 Additional 2HAM Definitions

Let α_0 be a producible supertile that grows into β via the supertile assembly sequence $\alpha_0, \alpha_1, \ldots$ and let δ be a producible supertile that can combine with α_0. Then β is *unfair* provided that, for every $i \geq 0$, δ can combine with α_i but does not. Otherwise, we say that β is *fair*. Note that if δ did combine with α, then the resulting supertile does not necessarily grow into β. Intuitively, if a supertile is able to bind to another growing supertile at any given step, it eventually does so if the latter is fair.

A shape is *aperiodic* provided there exists no non-trivial translation of the shape that yields itself. That is, the shape has no translational symmetries. A TAS is *aperiodic* provided every terminal supertile has an infinite aperiodic shape.

2.3 2HAM Variants

In the *two-handed restricted-glue Tile Assembly Model* or *rg2HAM*, glue strengths come from the set $\{-1, 0, 1\}$ and there is a unique glue of strength -1. Negative-strength glues permit producible supertiles α and β such that $\gamma = \alpha + \beta$ is *not* τ-stable. Producible supertiles that are not τ-stable can *break* into supertiles along cuts of strength less than τ. A supertile is *terminal* provided it cannot combine with any other producible supertile and cannot break.

In the *3D two-handed Tile Assembly Model* or *3D2HAM*, tiles are unit cubes. As in 2D, the tiles do not rotate, and each face of a tile has a glue.

3 Universal Computation in the rg2HAM

In this section, we prove that Turing-universal computation is possible in the temperature-1 2HAM when a single negative-strength glue is also permitted. Without such a glue, spurious nucleation would seemingly cause the vast majority of produced supertiles to nucleate as encodings of random configurations of the Turing machine and run nonsense computations both forward and backward. This uncountably large sea of undesired supertiles would then dilute supertiles encoding the desired computation.

Our construction, described in the proof of Theorem 1, utilizes small "jack-hammer" supertiles that break up the sea of undesired supertiles into constant-sized terminal supertiles (from a constant-sized set). This "jackhammering" is carried out on a "zig-zag" simulation of a modified version of the input Turing machine. Stated formally:

Theorem 1. *Let M be a Turing machine M with tape alphabet Γ. There exists an equivalent machine M'' and tile set T such that for any terminal supertile α of $\mathcal{T} = (T, 1)$ encoding a valid computation of $M''(x)$ and fair $\beta \in \mathcal{A}[\mathcal{T}]$, either β is a subassembly of α or β can be broken apart so that every subassembly of β can grow into a $O(\log |\Gamma|)$-sized terminal supertile.*

To prove Theorem 1, we first define an intermediate machine, M'. Let M' be a Turing machine which is equivalent to M but with the following modifications:

1. A new **end** symbol in the tape alphabet.
2. A new set of **head** symbols in the tape alphabet, one for each element of $\{*\} \times \Gamma$.
3. A new set of tape alphabet symbols that have additional markings to denote that the cells containing them occur to the right of the tape head (with cells to the left unmarked by these symbols). The new symbols are the elements of $\{R\} \times \Gamma$.
4. The assumption that the initial tape contains the input string x padded by one copy of **end** on each side, and with exactly one **head** symbol denoting the location of M's read/write head along with the value of the tape cell.

The tape alphabet of M' has size $3|\Gamma| + 1$. M' operates in a "zig-zag" manner, traversing the complete tape from left-to-right, then right-to-left, etc. Each *zig* (left-to-right) or *zag* (right-to-left) traversal carries out at most one step of M, doing so only if the traversal direction matches the direction moved by the head. The left- and rightmost ends of the tape are denoted by the *end* symbol; each time the *end* symbol is reached, the tape is extended by one cell via moving the *end* symbol.

During each zig, M' begins on the leftmost input character and moves right until encountering the **end** symbol. Initially, the leftmost tape cell value contains the **head** symbol and all other tape cells contain the R symbol denoting that they are to the right of the head. The transitions of M' are updated versions of those of M that output values that indicate which side of the head the cell lies on once the head moves. During traversal, the symbols of all cells that do not contain the **head** symbol or are immediately right of such a cell are left unchanged.

For the cell containing a **head** symbol (encoding also the tape symbol in M), if M performs a right-moving transition when in the start state and given the cell value there, then the value of that cell is updated to the output of that transition, the cell to the immediate right is updated to contain a **head** symbol, and the state of M' is updated to encode the new state of M reached after that transition. Cells not containing the head symbol are left unchanged. Identical but symmetric behavior is performed during each zag.

We now define another new Turing machine M'' equivalent to M' but with the following modifications. Let $b = \lceil \texttt{bin}(|\Gamma'|+1) \rceil$, where Γ' is the tape alphabet of M'. That is, b is the length of the binary representation of the size of $|\Gamma'|+1$. Since $|\Gamma'| = 3|\Gamma| + 1$, $b = O(\log(|\Gamma|))$. The tape alphabet for M'' is $\{0, 1\}$ and M'' simulates M' by representing each of the characters, or cell values, of M' as a binary string of length b, i.e. each element of Γ' will be assigned a unique b-bit binary string between 1 and $|\Gamma'|$, a new *pad* symbol is represented as $11 \ldots 1$, and the *end* symbol is represented as $00 \ldots 0$.

Also, M'' expects an input tape that encodes (using the previously described encoding) an input tape of M' in binary, but with the *pad* symbol inserted between every pair of cell values and starting with a *pad* symbol on the left side. M'' operates identically to M', but reads, writes, and moves using b steps each, due to encoding each cell of M' as b consecutive cells. As M'' reads a "cell" (b consecutive cells) it writes the *pad* string; M'' then writes the cell's new symbol in the adjoining *pad* string, avoiding simultaneous reading and writing of a cell. Thus after completing each zig or zag, the adjacent cell and *pad* locations alternate, seen in Fig. 1. In conclusion, M'' simulates M using a binary tape alphabet and a head that moves in a zig-zag manner.

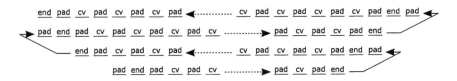

Fig. 1. A high-level overview of the growth and tape cell pattern created by $\mathcal{T}$ as it simulates $M(x)$, by simulating the machine M''. The *cv* elements are tape cells and *pad* elements are not.

Next, we create a tile set T such that the rg2HAM system $\mathcal{T} = (T, 1)$ has a unique terminal supertile with size larger than $O(b) = O(\log|\Gamma|)$. This terminal supertile contains an accurate computational history of M'' on the input x' (and thus M on input x) and so simulates $M(x)$. The tile set is the union of small groups of tiles that form functional supertiles called *gadgets*.

For the analysis of this system, we consider a specific "seeded" growth sequence where tiles attach one at a time, starting with a "seed" that grows into a "seed row" encoding the initial state of the machine, including an input tape. We prove that all other assembly sequences either result in the same terminal supertile (correctly representing the computation and containing the seed), or as "junk" supertiles of size $O(|\log \Gamma|)$.

Figure 2 shows the general structure of gadgets used to form the seed and zig rows. The first set of tiles to be created for T are those that form the seed row. A hard-coded set of gadgets is constructed that only bind to each other and in the correct pattern to encode the seed row: the appropriate binary encoding of x, interspersed by representations of *pad* and with encodings of *end* on the sides).

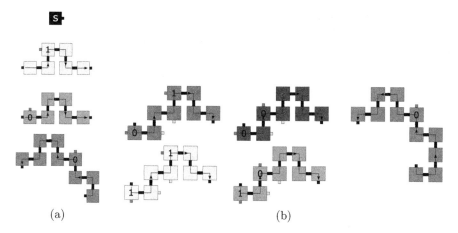

(a)   (b)

Fig. 2. (a) Seed row gadgets. (b) Zig row gadgets. Grey gadgets read bits on the bottom left, and write bits on the top left. Red tabs indicate -1-strength glues; black and yellow tabs indicate 1-strength glues. The gold gadget is the final gadget of a row and does not read but writes a 0 to the new row. Note that for compactness all gold gadgets as depicted write only one 0, but actually write a sequence of b 0's on the right (*end*), then b 1's to the left of those (**pad**), extending the previous row by $2b$ bits. Arrows only show the direction of growth if growth began from the seed. (Color figure online)

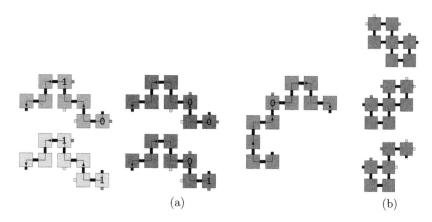

(a)   (b)

Fig. 3. (a) Zag row gadgets. Blue and green gadgets read bits on the bottom right, and write bits on the top right. The gold gadget is the final gadget of a row and does not read but writes a 0 in the new row. Note that arrows only show the direction of growth if growth began from the seed. (b) Jackhammer gadgets. The bottommost is a special type that attaches to a partial seed row that does not contain the seed. (Color figure online)

Next, we construct a similarly hard-coded zag row (Fig. 3a) that can only attach to the seed row and is also hard-coded. The next set of gadgets are copies of each of the types of gadgets for zig (Fig. 2b) and zag (Fig. 3a) rows that are specific to each state of M'', so that the current state of M'' is transmitted through the glues and the correct transitions are carried out when at *head* positions.

Incorrect growth is "jackhammered" apart by using yellow glues (see Figs. 4 and 5) to attach jackhammer and stopper gadgets to attach to the bottoms of rows. These gadgets break apart supertiles with bottom rows that are not the seed row: invalid computations and valid computations spuriously nucleated.

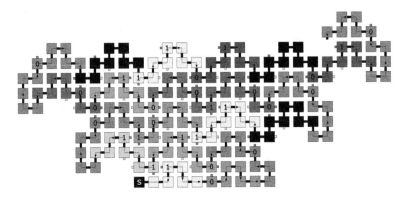

Fig. 4. An example supertile which has correctly grown five rows upward from the hard-coded seed (marked with "S"). The arrows show the direction of growth if growth began from the seed (possible but not necessary). (Color figure online)

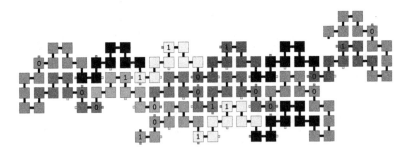

Fig. 5. A supertile which can be assembled but does not contain the hard-coded seed row. (Color figure online)

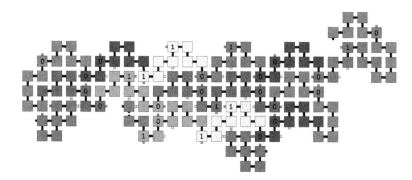

Fig. 6. A few of the locations where (red) jackhammer gadgets can attach. Note that the single-tile partial jackhammer (middle) is currently blocked from further growth. Further breakage by other jackhammers will eventually allow this jackhammer to complete. (Color figure online)

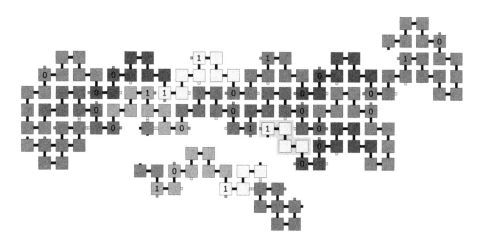

Fig. 7. (Bottom) The minimal supertile which can be separated by the rightmost jackhammer gadget. (Top) The tiles highlighted in yellow show all additional tiles that could have detached with it. Note that now the middle jackhammer is free to grow, and also that the leftmost jackhammer (and any which attach to gadgets at the end of rows) is not able to break off any supertile. (Color figure online)

Figures 6, 7 and 8 shows an example portion of such a supertile being broken apart into terminal junk supertiles. The resulting broken off pieces have size $O(\log |\Gamma|)$.

Figures 4 and 5 depict example supertiles that do and do not represent valid computations. Figure 6 shows an example of jackhammer gadgets attaching to a supertile that does not represent (part of) a valid computation.

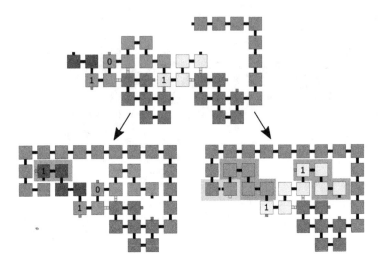

Fig. 8. (Top) The separated supertile from Fig. 7 with the attachment of another jack-hammer gadget (as well as a portion of a stopper gadget in pink) causing it to break into two pieces. On each of those pieces the stopper gadgets grow. (Left) Orange high-lighted portion shows partial regrowth of the row gadget, but the stopper gadget (pink) completes its growth, preventing any more of its growth. (Right) Orange highlighted portions show regrowth of row gadgets, stopper gadget growth is toward potential con-tention locations (yellow highlighted). If the stopper gadget grows into both of those locations first, the supertile becomes terminal, otherwise further row gadget growth creates a new location where another jackhammer can attach and split the supertile, eventually allowing the stopper gadget to block further growth. (Color figure online)

4   Efficient Square Assembly in the Temperature-1 3D2HAM

Here we augment the 2HAM with a third dimension, rather than negative-strength glues, and prove that the additional geometric freedom permits efficient assembly of squares that are one or two layers in the third dimension. More pre-cisely, the construction uses $O(\log^2 n)$ tile types to assemble terminal supertiles that all have a $n \times n \times 2$ bounding box and have a unique $n \times n$ projection into two dimensions. Let $\mathbb{N}_{n-1} = \{0, \ldots, n-1\}$. Formally stated:

Theorem 2. *There exists a tile set T with $|T| = O(\log^2 n)$ such that the 3D2HAM system $\mathcal{T} = (T, 1)$ self-assembles a shape S_n with $\mathbb{N}_{n-1}^2 \times \{0\} \subseteq S_n \subseteq \mathbb{N}_{n-1}^2 \times \{0, 1\}$.*

Proof. The construction begins with the temperature 1 3D aTAM counter of Cook, Fu, and Schweller [7]. This counter simulates the classic temperature-2 zig-zag style counter [17] (see also Sect. 3 by replacing cooperative binding with geometric blocking.

This counter fails in the 3D2HAM in two ways. First, all digits of the counter use a common constant-sized set of tiles. In the 3D2HAM, this enables counter rows of unbounded length to grow. Replacing the common tile set with distinct tile sets for each bit of the counter limits rows to exactly the desired quantity of $\Theta(\log n)$ bits.

The second failure is more fundamental: correct growth is only guaranteed when growth starts from the seed and proceeds forward. In the temperature-1 2HAM, assembly can spuriously nucleate at any tile type, allowing the counter to grow backwards from any row. This can lead to erroneous supertiles that increment incorrectly.

As a remedy, we use a *cyclic counter* consisting of two instances of the counter design of Fig. 10 that grow in opposite directions and initiate the growth of each other upon completion. See Fig. 9 for a schematic of the cyclic counter.

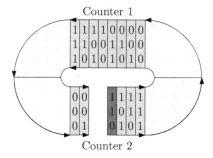

Fig. 9. The cyclic counter combines two forward growth counters that seed each other upon completion. Erroneous backward growth (red) halts, while forward growth proceeds until the crashing into the error to yield a full-length terminal supertile. (Color figure online)

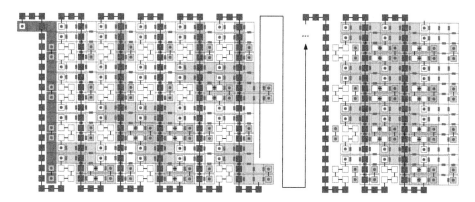

Fig. 10. Forward (left-to-right) growth occurs as in the counter of [7]. The counter value in each column is incremented by reading the geometry of the column to its left. White and grey blocks denote 0 and 1 bits, respectively. Smaller and larger squares denote tiles in the $z = 1$ and $z = 0$ layers, respectively.

Each counter is also modified to halt backwards growth when such growth makes an error. These ideas together resolve the problem of backwards growth, as failed backwards growth is eventually met by (correct) forward growth around the two-counter cycle that necessarily includes a complete correct counter. A fully formed counter is seen in Fig. 10. Due to space constraints, the details of how backward growth is halted are omitted.

Assembling $n \times n$ Squares. The cyclic counter counts correctly using $O(\log n)$ tile types, but only to values of n that are powers of 2. A counter for a non-power-of-2 value n involves concatenating up to $O(\log n)$ distinct cyclic counters, one for each 1-bit of the binary representation of n. Each counter uses $O(\log n)$ distinct tile types, and the counters are *padded* to a common width. Thus $n \times O(\log n)$ rectangular supertiles for arbitrary n can be assembled using $O(\log^2 n)$ tile types.

Such rectangles are assembled into squares by using one rectangle as a *backbone* and attaching additional rectangles to the backbone at regular $O(\log n)$ intervals. Such spacing can be achieved by modification of the counter to place a special glue based on the values of the least significant $\log \log n$ bits of the counter.   □

5   An Aperiodic 3D $\tau = 1$ 2HAM System

Here we describe an aperiodic system in the three-dimensional 2HAM at temperature 1, complementing the efficient square construction of the previous section.

Theorem 3. *There exists an aperiodic 3D2HAM system $T = (T, 1)$.*

Proof. Let the aTAM TAS $\mathcal{T}_{\text{count}} = (T_{\text{count}}, \sigma', 2)$ be the well-known system that assembles a "zig-zag" binary counter [17] and let $\alpha_{\text{term}} \in \mathcal{A}_\square[\mathcal{T}_{\text{count}}]$. Our system T assembles an infinite set of infinite terminal supertiles, each consisting of two reflected scaled versions of α_{term} and some extra "junk".

For every $\alpha \in \mathcal{A}[\mathcal{T}_{\text{count}}]$ there is a supertile in $\mathcal{A}[T]$ corresponding to α and composed of *macrotiles*: scaled versions of tiles in α where some tile glues are replaced by geometric "dents" and "bumps" encoding the glue's type in binary. Figure 11 shows how glues are replaced by "bit-reading" geometry.

Macrotiles bind to form arbitrarily long *strips* corresponding to binary counter rows in $\mathcal{T}_{\text{count}}$, including *end macrotiles* that match the tiles at the ends of each row. Two strips encoding adjacent binary values can attach vertically via glues on their end macrotiles and matching geometry (see Fig. 12).

As in the binary counter of Sect. 4, the forward growth of the counter is always correct. Thus any assembly in $\mathcal{A}[T]$ containing the macrotile corresponding to the seed of $\mathcal{T}_{\text{count}}$ corresponds to a subassembly of α_{term}.

Recall that the aim of T is to assemble only aperiodic terminal assemblies. Currently, T has (aperiodic) binary counter and (periodic) counter rows terminal supertiles. Dealing with the infinite periodic rows produced by the (necessarily) unlimited growth is the primary difficulty of the construction. As a solution, two mirrored copies of the counter tile set with identical behavior and disjoint glues

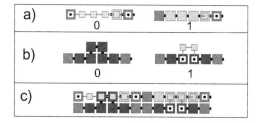

Fig. 11. (a) The bit-readers. Only one of them can assemble after "reading" a bit-writer. The olive-colored tiles attempt to grow both the aqua and yellow tiles but the geometry presented by the bit-writers prevent one of the paths from growing. (b) The two bit-writers. (c) An example of bit-readers "reading" two bit-writers. Smaller and larger squares denote tiles in the $z = 1$ and $z = 0$ layers, respectively. (Color figure online)

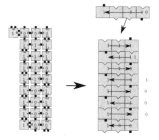

Fig. 12. A producible assembly of $\mathcal{T}_{\text{count}}$ and corresponding macrotile schematic diagram. The geometry of the macrotiles in adjacent strips ensure that adjacent rows encode incremented binary values.

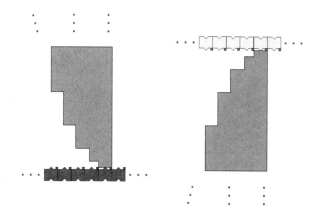

Fig. 13. The moderately shaded regions represent the scaled up macrotile version of the counter supertile and its reflection. The darkly shaded tiles are strips of tiles which expose a `maroon` glue and the lightly shaded tiles are strips of tiles which expose a `red` glue. (Color figure online)

are combined and modified to ensure infinite periodic supertiles of one attaching to the aperiodic supertiles of the other. Due to space constraints, we only sketch the implementation of this idea.

Let T_1 be the tile types described in $\mathcal{T}$ thus far and β_1 be the terminal supertile corresponding to α_{term} assembled from these tile types. Then $\mathcal{T} = (T_1 \cup T_2, 1)$, where T_2 are tile types with glues disjoint from those in T_1 that form a vertically reflected version of β_1 called β_2. Let S_1 be the macrotile in β_1 corresponding to the seed in $\mathcal{T}'_{\text{count}}$ and let S_2 be the counterpart of S_1 in β_2. A maroon glue is added to the south face of the southernmost tile of each macrotile in β_1, unless the macrotile corresponds to the seed of $\mathcal{T}'_{\text{count}}$, where a red glue is added. Similarly, a red glue is added to the north face of the northernmost tile of each macrotile in β_2, unless the macrotile corresponds to the seed of $\mathcal{T}_{\text{count}}$, where a maroon is added.

Any infinitely long strip is not terminal unless an infinite number of seeds (of the reflected system) bind to their red or maroon glues (see Fig. 13). Moreover, at least one of these seeds must be contained in an infinite counter. Thus every terminal supertile has β_1 or β_2 as a subassembly and so all terminal supertiles of $\mathcal{T}$ are aperiodic. □

References

1. Adleman, L., Cheng, Q., Goel, A., Huang, M.-D.: Running time and program size for self-assembled squares. In: Proceedings of the 33rd Annual ACM Symposium on Theory of Computing (STOC), pp. 740–748 (2001)
2. Barish, R.D., Schulman, R., Rothemund, P.W., Winfree, E.: An information-bearing seed for nucleating algorithmic self-assembly. Proc. Natl. Acad. Sci. **106**(15), 6054–6059 (2009)
3. Berger, R.: The undecidability of the domino problem. Mem. Am. Math. Soc. **66**, 1–72 (1966)
4. Cannon, S., Demaine, E.D., Demaine, M.L., Eisenstat, S., Patitz, M.J., Schweller, R., Summers, S.M., Winslow, A.: Two hands are better than one (up to constant factors): self-assembly in the 2HAM vs. aTAM. In: Proceedings of 30th International Symposium on Theoretical Aspects of Computer Science (STACS). LIPIcs, vol. 20, pp. 172–184. Schloss Dagstuhl (2013)
5. Chen, H.-L., Doty, D., Manuch, J., Rafiey, A., Stacho, L.: Pattern overlap implies runaway growth in hierarchical tile systems. In: Arge, L., Pach, J. (eds.) 31st International Symposium on Computational Geometry (SoCG). LIPIcs, vol. 34, pp. 360–373. Schloss Dagstuhl (2015)
6. Chen, H.-L., Schulman, R., Goel, A., Winfree, E.: Reducing facet nucleation during algorithmic self-assembly. Nano Lett. **7**(9), 2913–2919 (2007)
7. Cook, M., Fu, Y., Schweller, R.T.: Temperature 1 self-assembly: deterministic assembly in 3D and probabilistic assembly in 2D. In: Proceedings of the 22nd ACM-SIAM Symposium on Discrete Algorithms, SODA 2011, pp. 570–589 (2011)
8. Demaine, E.D., Demaine, M.L., Fekete, S.P., Ishaque, M., Rafalin, E., Schweller, R.T., Souvaine, D.L.: Staged self-assembly: nanomanufacture of arbitrary shapes with $O(1)$ glues. Nat. Comput. **7**(3), 347–370 (2008)
9. Doty, D.: Producibility in hierarchical self-assembly. Nat. Comput. **15**(1), 41–49 (2016)

10. Doty, D., Patitz, M.J., Summers, S.M.: Limitations of self-assembly at temperature 1. Theor. Comput. Sci. **412**, 145–158 (2011)
11. Fekete, S.P., Hendricks, J., Patitz, M.J., Rogers, T.A., Schweller, R.T.: Universal computation with arbitrary polyomino tiles in non-cooperative self-assembly. In: Proceedings of the 25th ACM-SIAM Symposium on Discrete Algorithms, SODA 2015, pp. 148–167. SIAM (2015)
12. Goodman-Strauss, C.: Open questions in tiling (2000). http://comp.uark.edu/strauss/papers/survey.pdf
13. Grünbaum, B., Shephard, G.C.: Tilings and Patterns. W.H. Freeman and Company, New York (1987)
14. Meunier, P.-E., Patitz, M.J., Summers, S.M., Theyssier, G., Woods, D.: Intrinsic universality in tile self-assembly requires cooperation. In: Proceedings of the 25th Symposium on Discrete Algorithms (SODA), pp. 752–771 (2014)
15. Padilla, J.E., et al.: Asynchronous signal passing for tile self-assembly: fuel efficient computation and efficient assembly of shapes. In: Mauri, G., Dennunzio, A., Manzoni, L., Porreca, A.E. (eds.) UCNC 2013. LNCS, vol. 7956, pp. 174–185. Springer, Heidelberg (2013)
16. Patitz, M.J., Schweller, R.T., Summers, S.M.: Exact shapes and turing universality at temperature 1 with a single negative glue. In: Cardelli, L., Shih, W. (eds.) DNA 17 2011. LNCS, vol. 6937, pp. 175–189. Springer, Heidelberg (2011)
17. Rothemund, P.W.K., Winfree, E.: The program-size complexity of self-assembled squares (extended abstract). In: Proceedings of the 32nd ACM Symposium on Theory of Computing (STOC), pp. 459–468 (2000)
18. Schulman, R.: The self-replication and evolution of DNA crystals. PhD thesis (2007)
19. Schulman, R., Winfree, E.: Synthesis of crystals with a programmable kinetic barrier to nucleation. Proc. Natl. Acad. Sci. **104**(39), 15236–15241 (2007)
20. Schulman, R., Winfree, E.: Programmable control of nucleation for algorithmic self-assembly. SIAM J. Comput. **39**(4), 1581–1616 (2009)
21. Seeman, N.C.: Nucleic-acid junctions and lattices. J. Theor. Biol. **99**, 237–247 (1982)
22. Socolar, J.E.S., Taylor, J.M.: An aperiodic hexagonal tile. J. Comb. Theor. Series A **118**(8), 2207–2231 (2011)
23. Soloveichik, D., Winfree, E.: Complexity of self-assembled shapes. SIAM J. Comput. **36**(6), 1544–1569 (2007)
24. Winfree, E.: Algorithmic self-assembly of DNA. PhD thesis, Caltech (1998)

Verifying Chemical Reaction Network Implementations: A Bisimulation Approach

Robert F. Johnson[1]([⊠]), Qing Dong[2], and Erik Winfree[1,3]

[1] Bioengineering, California Institute of Technology, Pasadena, USA
rfjohnso@caltech.edu
[2] Computer Science, SUNY Stony Brook, Stony Brook, NY, USA
[3] Computer Science, California Institute of Technology, Pasadena, USA

Abstract. Efforts in programming DNA and other biological molecules have recently focused on general schemes to physically implement arbitrary Chemical Reaction Networks. Errors in some of the proposed schemes have driven a desire for formal verification methods. We show that by interpreting each implementation species as a set of formal species, the concept of weak bisimulation can be adapted to CRNs in a way that agrees with an intuitive notion of a correct implementation. We give examples of how to use bisimulation to prove the correctness of an implementation or detect subtle problems. We examine the complexity of finding a valid interpretation between two CRNs if one exists, and that of checking whether an interpretation is valid. We show that both are PSPACE-complete in the general case, but are NP-complete and polynomial-time respectively under an assumption that holds in many practical cases. We give algorithms for both of those problems.

1 Introduction

In molecular programming, many real and abstract systems can be expressed in the language of Chemical Reaction Networks (CRNs). A CRN specifies a set of chemical species and the set of reactions those species can do, and the CRN model allows us to deduce the global behavior of the system from that local specification. CRNs are a useful way to separately analyze the computational and the physical aspects of a system. We can use the CRN model to help analyze real systems [3,4] or design engineered systems [5,17].

Despite this ideal, there remains a gap between abstract and real CRNs. To illustrate this gap, consider the approximate majority CRN [1,5]:

$$X + Y \xrightarrow{k} 2B$$

$$X + B \xrightarrow{k} 2X$$

$$Y + B \xrightarrow{k} 2Y$$

This abstract CRN quickly and with high probability converts all of the initial X and Y molecules into the same amount of whichever one was initially greater [1].

© Springer International Publishing Switzerland 2016
Y. Rondelez and D. Woods (Eds.): DNA 2016, LNCS 9818, pp. 114–134, 2016.
DOI: 10.1007/978-3-319-43994-5_8

However, no three molecules with exactly this behavior are known to exist. (In a strict sense, no three molecules with *exactly* this behavior can exist, because for all three reactions to be driven forward would require $X + Y$ to be both lower-energy and higher-energy than $2B$.) For contrast, consider the DNA strand displacement system built by Chen et al. [5] meant to implement this abstract CRN. The DNA system uses additional molecules which are consumed as "fuel" to drive these three reactions, ending up with over 25 each of species and reactions. Without knowing that it is meant to be an implementation of the approximate majority CRN, it might be difficult to tell what the DNA system was meant to do. Even knowing the correspondence, it is not obvious that there is no mistake in that complex implementation.

The issue of verifying correctness is exacerbated by the recent profusion of experimental and theoretical implementations in synthetic biology and molecular programming. Of particular interest to us, Soloveichik et al. [16] designed a systematic way to construct a DNA system to simulate an arbitrary CRN. Since then there have been a number of methods to translate an arbitrary CRN into a DNA strand displacement circuit [2,12,16]. While each one gave arguments for why it was a correct implementation, they did not come with a general theory of what it means to correctly implement a CRN. In some cases this led to subtle problems, of which we will give examples later. To be certain that such implementations are correct, CRN verification methods were invented. Such methods include Shin's pathway decomposition [15], Lakin et al.'s serializability analysis [9], and Cardelli's morphisms between CRNs [3].

We present a method for comparing an implementation CRN with an abstract CRN based on the concept of bisimulation from concurrency theory [11]. Our method associates each implementation species with a multiset of formal species, then asserts correctness if the reactions reachable from any implementation state are the same as the corresponding state in the abstract CRN. Like pathway decomposition [15] and serializability [9] but unlike Cardelli's morphisms [3], our bisimulation method works with the stochastic model for low-copy-number CRNs and doesn't take into account rates or kinetics. The use of interpretations instead of pathways means that some implementations considered correct by pathway decomposition are considered incorrect by bisimulation and vice versa. Interpretations also make bisimulation more local than pathway decomposition or serializability, which we hope makes it more understandable and tractable. We show how bisimulation can be used to prove a CRN implementation correct or identify subtle problems. We present an algorithm to check whether a particular interpretation between two CRNs is a bisimulation relation, and an algorithm to find such an interpretation if one exists. We analyze the computational complexity of both problems. We prove that both are PSPACE-complete in the general case but become polynomial time and NP-complete, respectively, when formal reactions are limited to a constant number of reactants. We hope this method can be used in both verifying that engineered systems match their specification and in comparing natural systems to a system simple enough to analyze.

2   The Chemical Reaction Network Model

We work within the *Chemical Reaction Network* (CRN) model. A CRN is a tuple $(\mathcal{S}, \mathcal{R})$, where $\mathcal{S}$ is a finite set of species and $\mathcal{R}$ a finite set of reactions. A *reaction* is itself a tuple (R, P), where the *reactants* R and *products* P are both multisets of species. We require that in any reaction $R \neq P$. We work with the stochastic model of CRNs, where the state of the system is represented by nonnegative integer counts of each species, and states transition discretely to other states when reactions occur. In particular, a state $S \in \mathbb{N}^{\mathcal{S}}$ can transition to a state T if there is some reaction $(R, P) \in \mathcal{R}$ such that $R \leq S$ and $S - R + P = T$. Often a probabilistic semantics is attached to this model, but for our purposes we only need to know whether something is possible.

We use the notation $\{|\ldots|\}$ for multisets interchangeably with the chemical notation, e.g. $2A + B$, $\{|A, A, B|\}$, and $\{|2A, B|\}$ all refer to the same state. Similarly, we sometimes use the chemical notation for reactions, e.g. $A + B \to 2C$ is the same as $(\{|A, B|\}, \{|2C|\})$. The "reversible reaction" notation $A + B \rightleftharpoons 2C$ is a shorthand for the two reactions $(\{|A, B|\}, \{|2C|\})$ and $(\{|2C|\}, \{|A, B|\})$. Multisets can be added and multiplied by scalars componentwise, and can be compared componentwise: $S \leq T \iff \forall_X S(X) \leq T(X)$, and $S < T$ if $S \leq T$ and $S \neq T$. If $S \leq T$ then subtraction $T - S$ is defined componentwise. Set operations involving multisets implicitly treat each multiset as the set of all objects which appear at least once; e.g. $\{|1, 1, 2|\} \subset \{1, 2, 3\}$ but $\{|1, 1, 2|\} \not\subset \{1\}$.

In this model, each possible behavior of a CRN is specified by a trajectory: an initial state $S_0 \in \mathbb{N}^{\mathcal{S}}$ together with a (finite or infinite) sequence of reactions $r_k = (R_k, P_k) \in \mathcal{R}$. A trajectory implicitly specifies a sequence of states $S_k = S_0 + \sum_{i \leq k}(P_k - R_k)$, but a sequence of states is not enough to specify a trajectory. For example, if $A \to B$ and $X + A \to X + B$ are both reactions, then the sequence of two states $(S_0, S_1) = (\{|X, A|\}, \{|X, B|\})$ does not specify which of those two reactions happened, which is sometimes important. A trajectory is *valid* if each reaction (R_k, P_k) can occur in the state resulting from the previous reactions; that is, $R_k \leq S_k$. In general when we speak of "the trajectories of a CRN" we mean the valid trajectories.

A state T is *reachable* from a state S if T is the result of a valid finite trajectory that starts in S. We say a state T is *coverable* from a state S if there is some $T' \geq T$ such that T' is reachable from S. While the set of reachable states (from any given initial state) is an important aspect of the behavior of a CRN, it does not contain all the information about that CRN. For example, the two CRNs $(\{A, B, C\}, \{A \to B, B \to C, C \to A\})$ and $(\{A, B, C\}, \{A \to C, C \to B, B \to A\})$ have exactly the same set of reachable states T from any given initial state S, but are clearly different in a meaningful way. If however the set of (valid) trajectories of two CRNs are the same, then the two CRNs must be identical: since in particular the length-zero trajectories (i.e. states) are the same, so the sets of species are the same, and the length-one trajectories (single reactions) are the same. We say that two CRNs are *isomorphic* if there is a bijection between the sets of species such that the set of reactions of one, after applying this bijection, equals the set of reactions of the other.

3   The Meaning of Correctness

3.1   Interpretations

Schemes for translating an arbitrary abstract CRN into a DNA Strand Displacement (DSD) implementation [2,12,16] provide designs for the necessary DNA molecules, but how these molecules interact is best described by a model of the relevant biophysics. Reaction enumerators such as Visual DSD [10] and Peppercorn [8] produce, given a set of DNA molecules, a description of their predicted interactions as a CRN, allowing us to compare it to the original CRN using the same language. We refer to the original abstract CRN as the *formal CRN* $(\mathcal{S}, \mathcal{R})$ and the model's enumerated CRN as the *implementation CRN* $(\mathcal{S}', \mathcal{R}')$, which is usually larger than the formal CRN. As a convention, we assume that the formal CRN and the implementation CRN make use of disjoint sets of species. (When using verification to compare a detailed model of a natural system with unknown function to a simpler abstract CRN with known function, the natural system is the implementation and the abstract system is the formal CRN.) There are two other important features typical of engineered implementation CRNs. First, there is typically for each formal species A an implementation species x_A intended to correspond specifically to it, sometimes called a "signal species". Second, certain implementation species must always be present for the system to work, and are designated "fuel species". Fuel species are typically assumed to be held at a constant concentration, for example by setting their initial concentration high enough that it does not vary significantly over the running time of the CRN. In this situation, we can approximate the implementation CRN by a simplified CRN with all fuel species removed; e.g. if g_1 is a fuel, the reaction $x_A + g_1 \to i_A$ can be replaced by $x_A \to i_A$ with no loss of meaning.

Figure 1 gives an example of this process for the formal reaction $A + B \to C + D$, yielding an implementation CRN with four reactions. (Names such as x_A

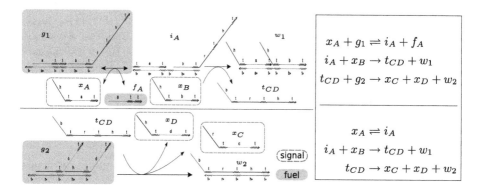

Fig. 1. Implementation of $A + B \to C + D$ using the scheme described in [16]. Left: DNA complexes and reactions. Top right: Direct translation of reactions in the implementation CRN. Bottom right: Implementation CRN after removing fuels.

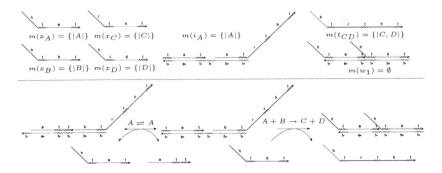

Fig. 2. Interpretation of the implementation CRN in Fig. 1. $m(t_{CD}) = A + B$ would also be a valid interpretation for this CRN.

and t_{CD} are based on the intent of the designers of the CRN, but the subscripts have no theoretical meaning.) The signal species x_A can freely convert to and from i_A, and the strand t_{CD} can produce the signals x_C and x_D (and waste w_2). Intuitively, i_A is an A and t_{CD} is a C and a D; in this sense the first and third reactions are silent, and the second is $A + B \to C + D$. We formalize this by defining an *interpretation* of the implementation species (Fig. 2):

Definition 1. *An* interpretation *is a function* $m : \mathcal{S}' \to \mathbb{N}^{\mathcal{S}}$ *from implementation species to multisets of formal species. We extend this linearly from species to states:* $m(\sum_{i=1}^{n} a_i X_i) = \sum_{i=1}^{n} a_i m(X_i)$. *We also define* $m(R' \to P') = m(R') \to m(P')$ *unless* $m(R') = m(P')$, *in which case* $m(R' \to P') = \tau$ *and we say the reaction is* trivial. *For example, if* $m(i_{AB}) = A+B$, $m(x_A) = A$, *and* $m(t_{BC}) = B+C$ *then* $m(i_{AB}+x_A) = 2A+B$, *and* $m(i_{AB} \to x_A+t_{BC}) = A+B \to A+B+C$.

The interpretation of an implementation reaction is always a pair (R, P) of multisets of formal species, or τ, but (R, P) may not be in $\mathcal{R}$. Any such pair is a *reaction in the language of the formal CRN*, but is a *formal reaction* only if $(R, P) \in \mathcal{R}$. Similarly, (R', P') is an *implementation reaction* only if it is in $\mathcal{R}'$.

In the following notation, S', T', S'', and T'' refer to implementation states; S and T to formal states; r' to an implementation reaction; and r to a reaction in the language of the formal CRN or τ. When a formal reaction r takes state S to state T, we write $S \xrightarrow{r} T$; $S' \xrightarrow{r'} T'$ is similar. Note that if $S \xrightarrow{r} T$, then $r = (R, P) \in \mathcal{R}$ as well as $S-R+P = T$, and analogously for the implementation. Further, we write $S' \xrightarrow{r} T'$ when $S' \xrightarrow{r'} T'$ for some r' with $m(r') = r$, which does not require $r \in \mathcal{R}$ (but does require $r' \in \mathcal{R}'$). Note that if $S' \xrightarrow{\tau} T'$ then $m(S') = m(T')$. To abstract away from trivial reactions, we write $S' \xRightarrow{\tau} T'$ to mean S' can reach T' via 0 or more trivial reactions, and $S' \xRightarrow{r'} T'$ when $S' \xRightarrow{\tau} S'' \xrightarrow{r'} T'' \xRightarrow{\tau} T'$. Note that $S' \xRightarrow{\tau} S'$ and $S \xRightarrow{\tau} S$ are always true. $S' \xRightarrow{r} T'$ for $r \neq \tau$ is again defined as $S' \xRightarrow{r'} T'$ for some r' with $m(r') = r$. $S \xRightarrow{r} T$ for $r \neq \tau$ is defined but trivial: $S \xRightarrow{r} T \iff S \xrightarrow{r} T$. When the final state is irrelevant, we sometimes write $S' \xRightarrow{r'}$, etc.

3.2   Three Notions of Correctness

Our notion of correctness is motivated by the earlier observation that the set of valid trajectories defines equivalence between formal CRNs, and allowing renaming of species defines isomorphism. Applying this notion to an implementation CRN with an interpretation introduces two difficulties. First, due to trivial reactions, the implementation trajectory may involve more steps. This is easily solved by defining the interpretation of an implementation trajectory to remove trivial reactions. Second, and more seriously, the full set of interpreted implementation trajectories may cover the formal trajectories, yet particular implementation trajectories may experience restricted options for alternative paths. An extreme example of this is an implementation CRN that is subject to deadlock, $(\{x_A, x_B, y_B, x_C\}, \{x_A \to x_B, x_A \to y_B, x_B \to x_C, x_C \to x_A\})$ with the interpretation $m = \{(x_A, A), (x_B, B), (y_B, B), (x_C, C)\}$, which has the same interpreted trajectories as the formal CRN $(\{A, B, C\}, \{A \to B, B \to C, C \to A\})$, which cannot deadlock. To resolve this issue, we need a finer-grained notion of trajectory equivalence that requires equivalence given any initial state. As defined formally below, this is a satisfactory definition of correctness.

However, since the sets of trajectories are generally infinite, we would like a more local definition that facilitates efficient computational analysis. We define three local conditions on the interpretation which we show are equivalent to trajectory equivalence. As further evidence that our notion of correctness is sound, we show that these three conditions are equivalent to a special case of weak bisimulation from concurrency theory [11]. This gives us three notions of correctness, given a formal CRN, an implementation CRN, and an interpretation:

I **Equivalence of trajectories**
  (i) The set of formal trajectories and interpretations of implementation trajectories are equal.
  (ii) For every implementation state S', the set of formal trajectories starting from $m(S')$ and interpretations of implementation trajectories starting from S' are equal.

II **Three conditions on the interpretation**
  (i) *Atomic condition:* For every formal species A, there exists an implementation species x_A such that $m(x_A) = \{|A|\}$.
  (ii) *Delimiting condition:* The interpretation of any implementation reaction is either trivial or a valid formal reaction.
  (iii) *Permissive condition:* If $S \xrightarrow{r}$ and $m(S') = S$, there exists an implementation reaction r' such that $m(r') = r$ and $S' \xRightarrow{r'}$.

III **Weak bisimulation**
  (i) For all implementation states S',
       if $S' \xrightarrow{r} T'$, then $S \xRightarrow{r} T$ where $S = m(S')$ and $T = m(T')$.
  (ii) For all formal states S, there exists S' with $m(S')=S$, and for all such S',
       if $S \xrightarrow{r} T$, then for some T', $S' \xRightarrow{r} T'$ and $m(T') = T$.

A few comments are in order. It may seem that the second condition for trajectory equivalence supercedes the first, but it does not: for example, the second condition may be satisfied even if there is no implementation state S' that is interpreted as formal state S, whereas the first condition will not be satisfied in that case.

Our definition of bisimulation in CRNs is in fact a special case of Milner's definition [11] for transition systems. Milner allows an arbitrary relation between states, while we rely on an interpretation m to establish a relation between formal states and implementation states. Our definition of the interpretation enforces several restrictions that, to us, are natural and consistent with the structure of CRNs: they provide a unique interpretation for each implementation state (i.e. the interpretation is a function), subsets of an implementation state can be interpreted separately and additively combined (i.e. the function is linear), and every formal state has at least one corresponding implementation state (i.e. the interpretation is surjective). In fact, *any* relation between formal states and implementation states that is a surjective linear function is induced by some interpretation, as shown in Lemma 1. Thus, we can take advantage of the finite specification of interpretations, while not losing any generality beyond the natural restrictions that we desire. These observations justify describing our notion of bisimulation in CRNs as "surjective linear weak bisimulation".

Lemma 1. *Let $\leftrightarrow \subset \mathbb{N}^S \times \mathbb{N}^{S'}$ be a relation between formal states and implementation states. If for every implementation state S' there is exactly one formal state S such that $S \leftrightarrow S'$ (function) and for every pair of pairs $S_1 \leftrightarrow S_1'$ and $S_2 \leftrightarrow S_2'$ we have $S_1 + S_2 \leftrightarrow S_1' + S_2'$ (linearity), then there is some interpretation $m : S' \rightarrow \mathbb{N}^S$ which, when extended to implementation states $m : \mathbb{N}^{S'} \rightarrow \mathbb{N}^S$, induces that relation: $S \leftrightarrow S' \iff S = m(S')$. Furthermore, for every S there is some S' such that $S \leftrightarrow S'$ (surjectivity) iff m satisfies the atomic condition.*

Proof. Given that the relation $\leftrightarrow$ is a linear function from $\mathbb{N}^{S'}$ to $\mathbb{N}^S$, we define the interpretation to be $m(x) = S_x$ where S_x is the unique formal state such that $S_x \leftrightarrow \{|x|\}$. Now, any implementation state S' is some sum of implementation species, $S' = \sum_{x \in S'} \alpha_x x$, and because we define the interpretation of a state as the sum of interpretations of species, $m(S') = \sum_{x \in S'} \alpha_x m(x)$. Then by the linearity assumption on $\leftrightarrow$, $m(S') \leftrightarrow S'$. Thus, if $S = m(S')$, then $S \leftrightarrow S'$. Conversely, if $S \leftrightarrow S'$, then $S = m(S')$ because $\leftrightarrow$ is a function.

If we further assume that $\leftrightarrow$ is surjective, then in particular for each formal species A, there must be some S' such that $\{|A|\} \leftrightarrow S'$, i.e. $m(S') = \{|A|\}$. Since $m(S')$ is the sum of interpretations of species in S' and an implementation species cannot interpret to fractional or negative formal species, there must be some species $x_A \in S'$ with $m(x_A) = \{|A|\}$ (and any other species in S' interpret to $\emptyset$). Thus the atomic condition is satisfied. Conversely, if the atomic condition is satisfied, then consider an arbitrary formal state $S = \sum_{A \in S} \alpha_A A$. Using linearity, let $S' = \sum_{A \in S} \alpha_A x_A$, so $m(S') = S$, and thus $\leftrightarrow$ must be surjective. $\square$

Theorem 1. *The three definitions of correctness, namely trajectory equivalence, the three conditions on the interpretation, and weak bisimulation, are equivalent.*

Proof. We show that trajectory equivalence implies the three conditions formulation; the three conditions imply weak bisimulation; and weak bisimulation implies trajectory equivalence.

Given trajectory equivalence, we prove the three conditions on m. First, for the atomic condition, consider applying condition I.(i) of trajectory equivalence to formal trajectories of length 0, which are just formal states, and in particular formal states $S_A = \{|A|\}$ for each formal species A. That the set of trajectories are equal implies that there is an implementation trajectory whose interpretation is the (zero-length trajectory) state S_A, i.e. an implementation state S'_A with $m(S'_A) = \{|A|\}$. Then as in Lemma 1, there is some species $x_A \in S'_A$ with $m(x_A) = \{|A|\}$, satisfying the atomic condition. For the delimiting condition, consider implementation trajectories of length 1, specifically for each implementation reaction $r' = (R', P')$ the trajectory $R' \xrightarrow{r'} P'$. If r' is trivial, that is $m(r') = \tau$, its interpreted trajectory is a zero-length trajectory; if not, its interpreted trajectory is $m(R') \xrightarrow{m(r')} m(P')$, which by trajectory equivalence must be a formal trajectory. For that to be so, $m(r')$ must be a reaction in $\mathcal{R}$, thus satisfying the delimiting condition. For the permissive condition, for every formal reaction $r = (R, P)$ and implementation state S' with $m(S') \geq R$, the trajectory $m(S') \xrightarrow{r} T$, where $T = m(S') - R + P$, is a formal trajectory. By condition I.(ii) of trajectory equivalence, there is an implementation trajectory starting in S' whose interpreted trajectory is $m(S') \xrightarrow{r} T$. (Note that condition I.(i) implies this for *some* S' with $m(S') = S$, but not necessarily for every S'.) To have that interpretation, that implementation trajectory must have some reaction r' with $m(r') = r$ and all other reactions trivial; this is the definition of $S' \xRightarrow{r}$, satisfying the permissive condition.

Given the three conditions, we prove weak bisimulation. Given any S' with $m(S') = S$ and $S' \xrightarrow{r'} T'$ where $r' = (R', P')$, by the delimiting condition either $m(r') = \tau$ is trivial or $m(r') = r = (R, P) \in \mathcal{R}$. If trivial, then $m(T') = m(S') = S$ and $S \xRightarrow{\tau} S$ is true by convention. If nontrivial, then $r \in \mathcal{R}$; since $S' \xrightarrow{r'}$ we must have $S' \geq R'$, thus $m(S') \geq m(R') = R$, and $S \xrightarrow{r} T$ (therefore $S \xRightarrow{r} T$) where $T = S - R + P$. Since $T' = S' - R' + P'$, $m(T') = m(S') - m(R') + m(P') = T$, satisfying condition III.(i) of weak bisimulation. Given any S, by Lemma 1 the atomic condition implies there exists an S' with $m(S') = S$. Given any such S' with $S \xrightarrow{r} T$ where $r = (R, P)$, by the permissive condition there is some r' with $m(r') = r$ and $S' \xRightarrow{r'}$, which is an abbreviation for $\exists_{T'} S' \xRightarrow{r'} T'$, which is further an abbreviation for $\exists_{S''} S' \xRightarrow{\tau} S'' \xrightarrow{r'} T'$. (Strictly speaking $S' \xRightarrow{r'} T'$ means there is some $S' \xRightarrow{\tau} S'' \xrightarrow{r'} T'' \xRightarrow{\tau} T'$, but since we are choosing an arbitrary T' we can take $T' = T''$.) Then $m(S') = m(S'') = S$ since they are connected by trivial reactions, and where $r' = (R', P')$ with $m(R') = R$ and $m(P') = P$ we have $T' = S'' - R' + P'$ so $m(T') = S - R + P = T$, satisfying condition III.(ii) of weak bisimulation.

Given weak bisimulation, we prove trajectory equivalence. We first prove condition I.(ii). Given any S_0' with $S_0 = m(S_0')$ and any implementation trajectory $(S_0', r_1', \ldots, r_k', \ldots)$ with $r_k' = (R_k', P_k')$, let $S_k' = S_{k-1}' - R_k' + P_k' = S_0' - \sum_{i \leq k} R_i' + \sum_{i \leq k} P_k'$. Letting $S_k = m(S_k')$ and $r_k = m(r_k')$, it follows that either $r_k = \tau$ or else $r_k = (R_k, P_k)$ and $S_k = S_{k-1} - R_k + P_k = S_0 - \sum_{i \leq k} R_k + \sum_{i \leq k} P_k$ by linearity of m. From bisimulation, since each $S_{k-1}' \xrightarrow{r_k'} S_k'$ we have either $r_k = \tau$ and $S_{k-1} = S_k$, or $r \neq \tau$ and $S_{k-1} \xrightarrow{r_k} S_k$, since for $r \neq \tau$ in the formal CRN $S \overset{r}{\Rightarrow} T \iff S \overset{r}{\to} T$. The interpretation of that implementation trajectory is exactly S_0 followed by those reactions $S_{k-1} \xrightarrow{r_k} S_k$ for which $r_k \neq \tau$, and thus the interpretation is a formal trajectory. Conversely, given S_0' with $S_0 = m(S_0')$ and any formal trajectory $(S_0, r_1, \ldots, r_k, \ldots)$ with $r_k = (R_k, P_k)$, letting $S_k = S_{k-1} - R_k + P_k = S_0 - \sum_{i \leq k} R_k + \sum_{i \leq k} P_k$, we construct an implementation trajectory whose interpretation is that formal trajectory. Given S_0', define inductively S_k' and r_k' to be an implementation state and reaction such that $S_{k-1}' \overset{r_k'}{\Rightarrow} S_k'$ with $m(r_k') = r_k$ and $m(S_k') = S_k$, which exists by condition III.(ii) of weak bisimulation. Expanding each $\overset{r_k'}{\Rightarrow}$ implicitly defines an implementation trajectory $(S_0', r_{1,1}'', \ldots, r_{1,l_1}'', r_1', r_{2,1}'', \ldots)$ where each $m(r_{k,j}'') = \tau$ and each $m(r_k') = r_k$; the interpretation of this trajectory is the formal trajectory $(S_0, r_1, \ldots, r_k, \ldots)$ as desired, proving condition I.(ii). Condition I.(i) follows from condition I.(ii) of trajectory equivalence and condition III.(ii) of weak bisimulation: every implementation trajectory starts from some S' and by condition I.(ii) its interpretation must be a formal trajectory starting from $m(S')$. Conversely, every formal trajectory starts from some S, by condition III.(ii) of weak bisimulation there is some S' with $m(S') = S$, and by condition I.(ii) of trajectory equivalence there is an implementation trajectory starting from S' whose interpretation is that formal trajectory.   □

3.3 Applying Bisimulation

We now consider how to use bisimulation to analyze our example implementation of $A + B \to C + D$. We use the three conditions formulation. The atomic condition is satisfied by the "signal species" x_A, x_B, x_C, and x_D. For the delimiting condition, we check each implementation reaction individually: $i_A + x_B \to t_{CD} + w_1$ is interpreted as $A + B \to C + D$, which is formal, while $x_A \rightleftharpoons i_A$ and $t_{CD} \to x_C + x_D + w_2$ are trivial. The permissive condition says that for every formal reaction and for every implementation state in which that reaction should be able to happen, it can. There is one formal reaction, $A + B \to C + D$, and any state in which it should be able to happen must contain an x_B and either an x_A or i_A, since those are the only species whose interpretations contain either B and/or A. If the state contains x_B and i_A, then the reaction $i_A + x_B \to t_{CD} + w_1$ can happen and satisfies the permissive condition. If the state contains x_B and x_A, then the trivial reaction $x_A \to i_A$ followed by $i_A + x_B \to t_{CD} + w_1$ satisfies the permissive condition.

Now consider a different case. Figure 3 shows an implementation of $A + B \rightarrow C + D$ as described by Qian et al. [12] as a means to implement stack machines, along with a natural interpretation. The species $i_{AB:CD}$ is interpreted as $C+D$, while $i_{A:BCD}$ is interpreted as A and x_B as B. This makes the reaction $i_{AB:CD} \rightarrow i_{A:BCD} + x_B$ interpreted as $C+D \rightarrow A+B$, which is not a valid formal reaction. Thus the delimiting condition is unsatisfied, and the implementation is not correct according to bisimulation; any other interpretation will have the same problem. This is not a problem for deterministic stack machines, but it does identify an error with this as a translation scheme for arbitrary CRNs: if the reaction $A+B \rightarrow C+D$ were put together with a reaction $C \rightarrow C+E$, then it would be possible to go from $\{|A, B|\}$ to $\{|A, B, E|\}$ in the implementation CRN when it is not possible in the formal CRN.

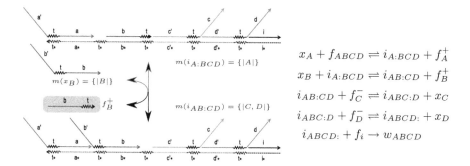

$$x_A + f_{ABCD} \rightleftharpoons i_{A:BCD} + f_A^+$$

$$x_B + i_{A:BCD} \rightleftharpoons i_{AB:CD} + f_B^+$$

$$i_{AB:CD} + f_C^- \rightleftharpoons i_{ABC:D} + x_C$$

$$i_{ABC:D} + f_D^- \rightleftharpoons i_{ABCD:} + x_D$$

$$i_{ABCD:} + f_i \rightarrow w_{ABCD}$$

Fig. 3. The translation scheme from [12], when used as a general CRN implementation, violates the delimiting condition. Species named f are fuels.

3.4   Properties of CRN Bisimulation

We describe two properties of CRN bisimulation that are likely to be useful in analyzing larger systems. While bisimulation in the classic sense is an equivalence relation between systems [11], our definition of interpretation-dependent CRN bisimulation is a partial order on the set of CRNs. In particular, CRN bisimulation is transitive, which allows us to do complex proofs of correctness in stages. We also show a modularity condition, where the combination of interpretations can be verified using only properties of each individual interpretation. This is particularly useful for general translation schemes where the translation of a whole CRN is the combination of one "module" for each reaction. As an example, we use modularity to prove that the translation scheme in [16] is correct for any CRN.

We first show that CRN bisimulation is transitive. Consider three CRNs: an abstract CRN $(\mathcal{S}, \mathcal{R})$, an implementation CRN $(\mathcal{S}'', \mathcal{R}'')$, and an intermediate CRN $(\mathcal{S}', \mathcal{R}')$. For example, $(\mathcal{S}, \mathcal{R})$ is an abstract CRN, $(\mathcal{S}'', \mathcal{R}'')$ is a low-level reaction enumeration of a prospective DNA implementation of $(\mathcal{S}, \mathcal{R})$, and $(\mathcal{S}', \mathcal{R}')$ is a more high-level reaction enumeration of the same DNA implementation which abstracts away from certain details. Say we have proven that $(\mathcal{S}', \mathcal{R}')$ is a valid implementation of $(\mathcal{S}, \mathcal{R})$ by finding an interpretation $m_1 : \mathcal{S}' \to \mathbb{N}^{\mathcal{S}}$ which is a bisimulation, and similarly have found an interpretation $m_2 : \mathcal{S}'' \to \mathbb{N}^{\mathcal{S}'}$ which is a bisimulation from $(\mathcal{S}'', \mathcal{R}'')$ to $(\mathcal{S}', \mathcal{R}')$. We want to prove that $(\mathcal{S}'', \mathcal{R}'')$, the system we actually have, is a valid implementation of $(\mathcal{S}, \mathcal{R})$, the system we want. The natural interpretation $m : \mathcal{S}'' \to \mathbb{N}^{\mathcal{S}}$ is $m(x) = (m_1 \circ m_2)(x) = m_1(m_2(x))$, treating m_2 as a function of species and m_1 as extended to a function of states. It turns out that this interpretation is in fact a bisimulation.

Lemma 2 *(Transitivity). If m_2 is a bisimulation from $(\mathcal{S}'', \mathcal{R}'')$ to $(\mathcal{S}', \mathcal{R}')$ and m_1 is a bisimulation from $(\mathcal{S}', \mathcal{R}')$ to $(\mathcal{S}, \mathcal{R})$, then $m = m_1 \circ m_2$ is a bisimulation from $(\mathcal{S}'', \mathcal{R}'')$ to $(\mathcal{S}, \mathcal{R})$.*

Proof. We use the three conditions formulation of correctness. We refer to $(\mathcal{S}, \mathcal{R})$ as the "formal" CRN, $(\mathcal{S}'', \mathcal{R}'')$ as the "implementation" CRN, and $(\mathcal{S}', \mathcal{R}')$ as the "intermediate" CRN. We show that each condition for m follows from the corresponding conditions for m_1 and m_2.

For any formal species A, by the atomic conditions for m_1 and m_2 there is an intermediate species x_A with $m_1(x_A) = \{|A|\}$ and implementation species y_A with $m_2(y_A) = x_A$. Then $m(y_A) = m_1(m_2(y_A)) = m_1(\{|x_A|\}) = \{|A|\}$, thus m satisfies the atomic condition.

For any implementation reaction $r'' = R'' \to P''$, by the delimiting condition for m_2 its interpretation $m_2(r'')$ is either an intermediate reaction $R' \to P' \in \mathcal{R}'$ or is τ. If $m_2(r'') = \tau$, that means $m_2(R'') = m_2(P'')$ and $m(R'') = m_1(m_2(R'')) = m_1(m_2(P'')) = m(P'')$, so $m(R'' \to P'') = m(r'') = \tau$. If $m_2(r'') = R' \to P'$ is a valid intermediate reaction, then $m(r'') = m_1(R' \to P')$, which by the delimiting condition for m_1 is either a valid formal reaction or trivial.

For any formal state S and reaction r with $S \xrightarrow{r}$ and any implementation state S'' with $m(S'') = S$, that means $S' = m_2(S'')$ is an intermediate state with $m_1(S') = S$. By the permissive condition on m_1, there is some r' with $m_1(r') = r$ and $S' \xRightarrow{r'}$. Using the permissive condition on m_2 and the argument used in Theorem 1 to show that the permissive condition implies trajectory equivalence, there is a sequence of implementation reactions starting from S'' which implements the intermediate trajectory by which $S' \xRightarrow{r'}$. This means that one of those reactions r'' has $m_2(r'') = r'$, some of them interpret via m_2 to various intermediate reactions in that pathway which are trivial under m_1, and the rest of which are trivial under m_2. An implementation reaction trivial under m_2 is trivial under m, as is a reaction which interprets under m_2 to an intermediate reaction trivial under m_1, thus all reactions in this pathway except r'' are trivial under m, so when viewed under m, $S'' \xRightarrow{r}$. $\qquad\square$

Bisimulation in the classic sense is an equivalence relation on states, which can be extended to an equivalence relation on systems [11]. Our definition of weak bisimulation introduces an asymmetry – one implementation state can only correspond to one formal state, but multiple implementation states can correspond to the same formal state. Bisimulation assumes a set of states and transitions between states where each transition is labelled from a common set of labels, while CRNs do not come with an obvious concept of labels. Our definition implicitly uses the set of all possible reactions using species in the formal CRN (plus the silent τ) as labels, labeling each formal reaction with itself and each implementation reaction with its interpretation. In that context, Lemma 1 and Theorem 1 say that there is an interpretation m which is a CRN bisimulation (satisfies the three conditions, has trajectory equivalence) if and only if there is a relation between states of that system which is a surjective and linear function from implementation states to formal states and is a bisimulation in the sense of [11]. Since we require one CRN to be designated the "formal" CRN in order to define a set of labels, it is difficult to make the concept of a symmetric relation between CRNs meaningful. Instead, CRN bisimulation is an order relation (up to isomorphism):

Lemma 3 *(Partial order). The following relation is a partial order: $(S', R') \gtrsim (S, R)$ if there exists an $m : S' \to \mathbb{N}^S$ which satisfies the atomic, delimiting, and permissive conditions (equivalently, its extension $m : \mathbb{N}^{S'} \to \mathbb{N}^S$ is a surjective linear weak bisimulation) with equality defined as $(S', R') \equiv (S, R)$ if there exists a bijection $n : S' \to S$ such that $(n(S'), n(R')) = (S, R)$ where n is extended naturally to sets and reactions.*

Proof. A partial order must be transitive, reflexive, anti-symmetric. Transitivity (if $a \leq b$ and $b \leq c$ then $a \leq c$) follows immediately from Lemma 2. Relexivity ($a \leq a$) is obvious by letting m be the identity function. It remains to show anti-symmetry (if $a \leq b$ and $b \leq a$, then $a = b$), i.e. that given (S_1, R_1) and (S_2, R_2) with $m_1 : S_1 \to \mathbb{N}^{S_2}$ and $m_2 : S_2 \to \mathbb{N}^{S_1}$ that each satisfy the atomic, delimiting, and permissive conditions, (S_1, R_1) and (S_2, R_2) are identical up to a change of species names. The atomic condition implies that $|S_1| \leq |S_2|$ and $|S_2| \leq |S_1|$, thus the numbers of species are equal and in particular m_1 is a bijection from species in S_1 to sets of exactly one species in S_2 (and the same is true for m_2). To simplify notation, we let $n(x) = y$ if $m_1(x) = \{|y|\}$; n must be a bijection from S_1 to S_2. (If the CRN has sufficient symmetry, it is not necessarily true that $m_2(n(x)) = \{|x|\}$, for example if both CRNs are $\{A \to C, B \to C\}$ we could have $m_2(n(A)) = \{|B|\}$.) Since n is a bijection, any reaction that would be trivial after interpretation (by either m_1 or m_2) must be trivial before interpretation, and thus cannot exist. By the delimiting condition for m_1, every reaction in R_1 must have its image under n in R_2; by the permissive condition for m_1, every reaction in R_2 must have its preimage under n in R_1; thus the two CRNs are equal up to the isomorphism n. □

In Sect. 3.3 we showed that the translation scheme from [16] is a correct implementation of the single reaction $A + B \to C + D$ according to CRN bisimulation. Intuitively, given a CRN of multiple reactions we should be able to combine the implementations of each such reaction to form a correct implementation of the CRN. In particular, we would like to show that the combined implementation CRN is correct using a condition which we can check on each individual reaction's implementation without having to check any property of the combined CRN. Since, as we will see in Sect. 4, the time required to check an interpretation scales much worse than linearly in the size of the implementation CRN, such a modularity condition would be a significant saving in the time required. While it is not in general true that combining two correct implementation CRNs gives a correct implementation of the combined formal CRN, there is a modularity condition which guarantees that the combined CRN is correct.

We consider an implementation CRN $(\mathcal{S}'_1, \mathcal{R}'_1)$ and formal CRN $(\mathcal{S}_1, \mathcal{R}_1)$ with interpretation $m_1 : \mathcal{S}'_1 \to \mathbb{N}^{\mathcal{S}_1}$, and another implementation CRN $(\mathcal{S}'_2, \mathcal{R}'_2)$ and formal CRN $(\mathcal{S}_2, \mathcal{R}_2)$ with interpretation $m_2 : \mathcal{S}'_2 \to \mathbb{N}^{\mathcal{S}_2}$, where both m_1 and m_2 are bisimulations. We assume the interpretations are compatible: for each $x \in \mathcal{S}'_1 \cap \mathcal{S}'_2$, $m_1(x) = m_2(x)$, which implies $m_1(x) \in \mathbb{N}^{\mathcal{S}_1 \cap \mathcal{S}_2}$. We also assume that the reactions in $\mathcal{R}'_1$ and $\mathcal{R}'_2$ are the only reactions that occur when you combine the implementation species in $\mathcal{S}'_1$ and $\mathcal{S}'_2$; that is, we assume no crosstalk reactions. Whether there is crosstalk can be checked by a reaction enumerator [8,10]. Aside from crosstalk, the main reason for the combined implementation to be incorrect according to bisimulation is some implementation species y in e.g. $\mathcal{S}'_1$ but not in $\mathcal{S}'_2$ whose interpretation contains a formal species $A \in \mathcal{S}_1 \cap \mathcal{S}_2$, where some formal reaction in $\mathcal{R}_2$ with A as a reactant cannot be implemented from an implementation state where y is the representation of A. If any such species y can, via trivial reactions, "release" any formal species in $\mathcal{S}_1 \cap \mathcal{S}_2$ in its interpretation to implementation species in $\mathcal{S}'_1 \cap \mathcal{S}'_2$, then we would think this problem cannot arise. This condition can be checked individually on each module without checking the combined CRN, and we show that this condition guarantees that the combined implementation is correct according to bisimulation.

Theorem 2 *(Modularity). Let m_1 be a bisimulation from $(\mathcal{S}'_1, \mathcal{R}'_1)$ to $(\mathcal{S}_1, \mathcal{R}_1)$ and m_2 be a bisimulation from $(\mathcal{S}'_2, \mathcal{R}'_2)$ to $(\mathcal{S}_2, \mathcal{R}_2)$ where m_1 and m_2 agree on $\mathcal{S}'_1 \cap \mathcal{S}'_2$. Let $\mathcal{S}' = \mathcal{S}'_1 \cup \mathcal{S}'_2$, $\mathcal{R}' = \mathcal{R}'_1 \cup \mathcal{R}'_2$, $\mathcal{S} = \mathcal{S}_1 \cup \mathcal{S}_2$, and $\mathcal{R} = \mathcal{R}_1 \cup \mathcal{R}_2$, and $m : \mathcal{S}' \to \mathbb{N}^{\mathcal{S}}$ equal m_1 on $\mathcal{S}'_1$ and m_2 on $\mathcal{S}'_2$. If for any $x \in \mathcal{S}'$ there is a sequence of trivial reactions $x \xRightarrow{\tau} Y + Z$ for $Y \in \mathbb{N}^{\mathcal{S}'_1 \cap \mathcal{S}'_2}$ and $m(Z) \cap (\mathcal{S}_1 \cap \mathcal{S}_2) = \emptyset$, then m is a bisimulation from $(\mathcal{S}', \mathcal{R}')$ to $(\mathcal{S}, \mathcal{R})$.*

Proof. We use the three conditions formulation. The atomic condition for m for each formal species A is satisfied by the species x_A that satisfy it for m_1 or m_2, as appropriate, or possibly both; e.g. if $A \in \mathcal{S}_1$ then there is some species $x_A \in \mathcal{S}'_1$ such that $m_1(x_A) = \{|A|\}$, which implies that $x_A \in \mathcal{S}'$ and $m(x_A) = m_1(x_A) = \{|A|\}$. Similarly the delimiting condition for m follows from that for m_1 and m_2: for any implementation reaction $R' \to P'$ in $\mathcal{S}'$, that reaction is in either $\mathcal{R}'_1$ or $\mathcal{R}'_2$, and its interpretation in m agrees with its interpretation in either m_1 or m_2 as appropriate, which is either a trivial reaction or a formal reaction in $\mathcal{R}_1$ or $\mathcal{R}_2$, which is thus in $\mathcal{R}$. (The delimiting condition assumes, as we mentioned above, that no crosstalk reactions exist, which when applying this theory to DNA implementations would be checked by a reaction enumerator.)

For the permissive condition, consider a formal reaction $r = R \rightarrow P$ and implementation state S' where $R \leq m(S')$. Either $r \in \mathcal{R}_1$ or $r \in \mathcal{R}_2$; without loss of generality say $r \in \mathcal{R}_1$. Divide S' into species in the first CRN and species not: let $S' = S_1' + S_2'$, where $S_1' \subset \mathcal{S}_1'$ and $S_2' \cap \mathcal{S}_1' = \emptyset$. If $m(S_1') \geq R$, then the permissive condition for m_1 applied to reaction r and state S_1' mean $S_1' \overset{r}{\Rightarrow}$, thus $S' \overset{r}{\Rightarrow}$ by the same sequence of reactions ignoring species in S_2'. In the general case, this means the proof is nontrivial only for formal species in R whose implementations in S' are in S_2', and we need to show that those formal species can be "extracted" into an implementation species in S_1'. This is exactly the modularity condition: for each species $x_i \in S_2'$ there is a sequence of trivial reactions by which $x_i \overset{\tau}{\Rightarrow} Y_i + Z_i$, where $Y_i \subset \mathcal{S}_1'$ and $m(Z_i) \cap S_1 = \emptyset$. In particular, since $R \rightarrow P$ is a reaction in CRN 1, $R \subset \mathcal{S}_1$ and $R \cap m(Z_i) = \emptyset$. We then have $S' \overset{\tau}{\Rightarrow} S_1' + Y + Z$, where $Y = \sum_i Y_i \subset \mathcal{S}_1'$ and $Z = \sum_i Z_i$. Since $R \cap Z = \emptyset$, $R \leq m(S')$, and $m(S') = m(S_1' + Y) + m(Z)$, we have $R \leq m(S_1' + Y)$. Since $S_1' + Y \subset \mathcal{S}_1'$, the permissive condition for m_1 implies $S_1' + Y \overset{r}{\Rightarrow}$, thus $S' \overset{r}{\Rightarrow}$. $\square$

DNA implementation schemes for arbitrary CRNs such as [2,12,16] typically have a set of common species and for each formal reaction a "module" with additional species and implementation reactions that implement the formal reaction. If the modules have no crosstalk and each one correctly implements its reaction and satisfies the modularity condition, then repeated applications of Theorem 2 prove that the entire CRN is a correct implementation.

Corollary 1. *Consider a formal CRN* $(\mathcal{S}, \mathcal{R})$ *with* n *reactions* $\mathcal{R} = \{r_i\}_{i=1}^n$, *and* n *implementation "module" CRNs* $(S_0' \cup S_i', \mathcal{R}_i')$ *with species* S_0' *in common, where any* S_i' *is disjoint from any* S_j' *for* $j \neq i$. *If there are interpretations* $m_i : S_i' \rightarrow \mathcal{S}$ *for* $0 \leq i \leq n$ *such that the interpretation* $(m_0 \cup m_i)$ *is a bisimulation from* $(S_0' \cup S_i', \mathcal{R}_i')$ *to* $(\mathcal{S}, \{r_i\})$, *and any* $x \in S_i'$ *can be converted* $x \overset{\tau}{\Rightarrow} Y + Z$ *by trival reactions in* $\mathcal{R}_i'$ *where* $Y \in \mathbb{N}^{S_0'}$ *and* $m(Z) = \emptyset$, *where* $m = \bigcup_{i=1}^n m_i$ *is the combination of the interpretations, then* m *is a bisimulation from* $(S_0' \cup \bigcup_{i=1}^n S_i', \bigcup_{i=1}^n \mathcal{R}_i')$ *to* $(\mathcal{S}, \mathcal{R})$.

In particular, the translation scheme from [16] discussed earlier satisfies the condition in Corollary 1 for $S_0' = \{x_A \mid A \in \mathcal{S}\}$, i.e. the signal species. Thus Corollary 1 proves that for any number of formal reactions, the scheme in [16] produces a correct implementation CRN, as long as the DSD reaction enumerator produces exactly the described reactions and no additional crosstalk reactions.

4   Checking Bisimulation

We now have a definition of "correct implementation", and can sometimes prove that a particular implementation is or is not correct. We would like to find a general way to check whether any implementation is correct.

We divide "checking bisimulation" into three questions. First, given a formal and implementation CRN and an interpretation, is the interpretation a bisimulation? Second, if (as in most engineered CRN implementations) we have a formal CRN, implementation CRN, and for each formal species A a designated signal species x_A, is there an interpretation which is a bisimulation and has $m(x_A) = \{|A|\}$? Finally, given a formal CRN, implementation CRN, and no additional information, is there an interpretation which is a bisimulation?

We give the complexity in terms of two parameters: the *size* n, the total number of species and reactions in the two CRNs, and the *arity* k, the maximum number of reactants in any formal reaction. We find the problem is easier when k is bounded by a constant, such as $k = 2$ limiting the formal CRN to bimolecular reactions.

4.1   Checking an Interpretation

First we consider the problem of, given an interpretation, checking whether it is a bisimulation. We use the three conditions on an interpretation, having proved in Theorem 1 that they are equivalent to bisimulation and trajectory equivalence. Given two CRNs and an interpretation between them, the atomic and delimiting conditions are trivial to check. This leaves only the permissive condition.

Checking the permissive condition means, for each formal reaction $r = (R, P)$ and implementation state S' with $m(S') \geq R$, S' can reach via trivial reactions some state from which a reaction that is interpreted as r can happen. Although there are infinitely many such S', we only need to consider the minimal such states. Consider two such states S' and S'' where $S'' > S'$ and $m(S'') \geq m(S') \geq R$. If there is some sequence of reactions by which $S' \overset{r}{\Rightarrow}$, then it can happen in S'' also and thus $S'' \overset{r}{\Rightarrow}$. If not, then the permissive condition is false for S', and we do not need to check S''. Thus we need only to check states S' such that $m(S') \geq R$ and there is no $S'_0 < S'$ for which $m(S'_0) \geq R$, which we call *minimal* implementation states (with respect to the given formal reaction r). All such minimal states can be enumerated by, for each reactant X_i in R, choosing some implementation species x_i such that $X_i \in m(x_i)$, removing $m(x_i)$ from R (ignoring any species in $m(x_i)$ not present in R), then applying this process recursively.

Now we have reduced the permissive condition to a finite problem: for each minimal S', check whether it can reach via trivial reactions some state T' from which a reaction r' with $m(r') = r$ can happen. By taking the reactants of each such r' in the implementation CRN, we have a list of multisets R' such that if for each minimal S' there is some such R' such that S' can reach a state greater than or equal to R', then the permissive condition is satisfied. This sounds similar to the covering (or superset reachability) problem: given states S' and T', can S' reach any state $T'' \geq T'$? Unfortunately, the covering problem was proven by Rackoff to be EXPSPACE-complete [13]. In particular, the covering problem is hard because to reach a given state T' from S' may require production and consumption of a large number of some species.

To solve the permissive condition in less than exponential space, we use the fact that for the permissive condition to be satisfied, we need a path from *every* minimal S' to some r' which implements r. Thus, if a minimal state $S'_0 \overset{\tau}{\rightarrow} S''$

for $S'' > S_1'$ which is minimal with $m(S_1') \geq R$, then either $S_1' \not\xrightarrow{r}$ and the permissive condition is false anyway, or $S_1' \xrightarrow{r}$ and we can treat $S_0' \xrightarrow{r} S''$ as $S_0' \xrightarrow{r} S_1'$ ignoring the extra species. Following this logic out, we visualize the state space as a graph of minimal states with an arrow from S_i' to S_j' if there is a trivial reaction $S_i' \xrightarrow{\tau} S_j' + \ldots$. We find that we can check the permissive condition using only paths through this graph with some loops of the form $S' \xrightarrow{\tau} S' + Z$ for some minimal state S' and multiset Z, which since trivial reactions do not change the interpretation implies that $m(Z) = \emptyset$, thus $m(y) = \emptyset$ for each $y \in Z$. If such a loop exists, then we know that arbitrarily many copies of each such y can be produced in state S', and we can ignore y whenever it appears as a reactant later on the path.

Lemma 4. *Let $(\mathcal{S}, \mathcal{R})$ and $(\mathcal{S}', \mathcal{R}')$ be a formal CRN and an implementation CRN, with interpretation m. Let $r = (R, P) \in \mathcal{R}$ be a formal reaction and S_0' an implementation state minimal for $m(S_0') \geq R$. Let z be the number of null species. If the permissive condition is satisfied, then there exists a sequence of $l \leq z$ multisets S_i' that are minimal for $m(S_i') \geq R$, l disjoint sets Z_i of null species, and nonnegative integers $(\alpha_i)_{1 \leq i \leq l+1}$, $(\beta_i)_{1 \leq i \leq l}$ such that, where $Y_i = \bigcup_{j \leq i} Z_j$, for each $1 \leq i \leq l$ there is a sequence of trivial reactions by which $S_{i-1}' + \alpha_i Y_{i-1} \xrightarrow{\tau} S_i'$ and $S_i' + \beta_i Y_{i-1} \xrightarrow{\tau} S_i' + Z_i$, and a sequence of trivial reactions by which $S_l' + \alpha_{l+1} Y_l \xrightarrow{r}$, where the same minimal implementation state is never covered twice within the same sequence. Conversely, if such paths exist for every formal reaction and minimal implementation state, then the permissive condition is satisfied.*

Proof (Sketch). If the permissive condition is satisfied, then for each minimal S', consider the first reaction on the shortest path by which $S' \xrightarrow{r}$. Starting from any given S_0', the pathway which at each S' takes that first reaction is a valid pathway, and either eventually implements r or eventually repeats the same minimal state S_1' twice. If it eventually implements r, then the pathway matches the desired pathway with all sequences of trivial reactions except the last one empty. If it eventually repeats, then for each reaction to be the first reaction on the shortest path, the sequence of reactions which loops must be $S_1' \xrightarrow{\tau} S_1' + Z_1$ for Z_1 a nonempty multiset of null species. With the sequence by which $S_0' \xrightarrow{\tau} S_1'$ and $S_1' \xrightarrow{\tau} S_1' + Z_1$ as the first two sequences of trivial reactions, consider a modified implementation CRN with all species in Z_1 removed; if the original implementation CRN satisfies the permissive condition, then making reactions easier cannot make the permissive condition false. Applying this by induction on the number of null species gives the remaining segments.

If such paths exist for a given formal reaction r and minimal implementation state S_0', then $S_0' \xrightarrow{r}$: from S_0' reach S_1'; produce "as many copies as needed" of Z_1 in the loop $S_1' \xrightarrow{\tau} S_1' + Z_1$, reach S_2', produce "as many copies as needed" of Z_2, etc. If such paths exist for every formal reaction r and minimal state S_0' for r, then every minimal state $S_0' \xrightarrow{r}$, thus as discussed above every state with $m(S') \xrightarrow{r}$ has $S' \xrightarrow{r}$, thus satisfying the permissive condition. $\square$

We describe two algorithms to check the permissive condition. One runs in space poly(nk) and time poly(n^{kn}); the other runs in space and time poly(n^k).

The space-efficient, loopsearch algorithm goes through each formal reaction (R, P) and each minimal implementation state S' in which $m(S') \geq R$ and searches for the path described in Lemma 4. It iterates through each partition of null species into sets Z_i and choice of states S'_i for which $S'_i \overset{\tau}{\Rightarrow} S'_i + Z_i$. A nonrepeating path from S'_i to S'_j or $S'_i + Z_i$ will have length at most N, where N is the number of minimal implementation states. Where $l = \lceil \log_2 N \rceil$, Savitch's theorem [14] says that a path from S'_a to S'_b of length at most 2^l can be found by checking all possible states S'_c for a path from S'_a to S'_c and S'_c to S'_b each of length at most 2^{l-1}, which can be done recursively. This algorithm stores at most $l + z$ minimal states plus a partition of z species at any one time.

The more time-efficient, graph-updating algorithm, for each formal reaction $r = (R, P)$ iteratively builds a table of minimal implementation states S'_i with $m(S'_i) \geq R$ and, for each minimal S'_i, which other minimal S'_j can be reached from S'_i via trivial reactions and which null species can be produced in a loop from S'_i to itself. In each iteration, for each S'_i that is not yet known to be able to implement r, for each trivial reaction of the form $S'_i + Z_1 \overset{\tau}{\rightarrow} S'_j + Y + Z_2$, where Z_1 and Z_2 contain only null species and all species in Z_1 are known to be producible in a loop from S'_i to itself, it updates the table as follows:

(i) If S'_j is known to be able to implement r, then S'_i can implement r. Otherwise:
(ii) For each $k \neq i$, if S'_j can reach S'_k, then S'_i can reach S'_k.
(iii) If S'_j can reach S'_i, then S'_i can produce in a loop any null species in Z_2, as well as any null species producible in a loop at S'_j.

The algorithm terminates when an iteration passes with no change to the table. If all states are known to be able to implement r, then the permissive condition is satisfied for r; otherwise the permissive condition is false. Using similar but slightly different reasoning as Lemma 4, we can prove that if the permissive condition is true, the algorithm will prove it in at most $(2zn^k + 1)n^k$ iterations.

Theorem 3. *Whether an interpretation is a bisimulation can be checked in polynomial space.*

Proof. The loopsearch algorithm takes polynomial space.   □

Theorem 4. *When the number of reactants in a formal reaction k is constant, whether an interpretation is a bisimulation can be checked in polynomial time.*

Proof (Sketch). We show that if the permissive condition is true, the graph-updating algorithm will prove it in at most $(2zn^k + 1)n^k$ iterations. Given a formal reaction $r = (R, P)$ and all states S' which are minimal for $m(S') \geq R$, at any given iteration for some S' and y with $m(y) = \emptyset$ it will be known that $S' \overset{\tau}{\Rightarrow} S' + y$. If the permissive condition is true, then as in the proof of Lemma 4, for each S' consider the first reaction on the shortest path by which $S' \overset{r}{\Rightarrow}$

assuming that each S_i' has infinite copies of any species y for which it is known that $S_i' \overset{\tau}{\Rightarrow} S_i' + y$. As in that proof, these reactions either give every S' a direct path by which $S' \overset{\tau}{\Rightarrow}$, or have at least one loop by which some $S_i' \overset{\tau}{\rightarrow} S_j' + y$ and $S_j' \overset{\tau}{\Rightarrow} S_i'$ which is not already known. (If not, then at least one of the reactions involved does not lead to the shortest path.) Let $l \leq n^k$ be the number of states in that loop. If every minimal state can implement r using only known null species, then the shortest paths each have length at most n^k, and algorithm will prove this and terminate in at most n^k iterations. If there is such a loop, it will take at most l iterations to prove that $S_i' \overset{\tau}{\Rightarrow} S_i' + y$, and an additional l iterations to prove for each other S_j' in that loop that $S_j' \overset{\tau}{\Rightarrow} S_j' + y$. Thus in at most $2n^k$ iterations one more fact will be known. The number of such facts is at most zn^k, all possible pairs of minimal state S' and null species y. If the permissive condition is true, it will be proven in at most $(2zn^k + 1)n^k$ iterations.   □

Although polynomial space in the general case is inefficient, we cannot do better. If we have (order of) n^k minimal states, it is possible to embed a PSPACE-complete computation in those n^k states. In particular, a Linear Bounded Automaton computation – a model of a Turing machine with space bounded by the size of its input, for which acceptance is a PSPACE-complete problem [7] – can be embedded in a polynomial-size implementation CRN, such that a given formal reaction is reachable in the implementation if and only if the LBA accepts.

Theorem 5. *CRN bisimulation in the general case is PSPACE-complete.*

Proof (Sketch). Consider a formal CRN with one reaction, $Q + A_1 + \ldots + A_n \rightarrow H$. An implementation CRN can simulate an arbitrary LBA with species representing the states of the LBA interpreted as Q and species representing the ith tape symbol interpreted as A_i. (For example, the reaction $q_i^0 + 0_i \rightarrow q_{i+1}^3 + 1_i$ for each $1 \leq i \leq n$ represents the Turing machine instruction, "in state 0, read a 0, write a 1, move right, go to state 3".) If the interpretation CRN can reset at any time to the starting state with the tape reading a given string x, and can implement the formal reaction only from an accepting state, then the permissive condition is true if and only if the LBA accepts the string x.   □

4.2   Finding an Interpretation

We now consider the problem of, given a formal and implementation CRN, can we find an interpretation that is a bisimulation or correctly assert that none exists? An algorithm to enumerate interpretations that satisfy the delimiting condition was given in [6]. This algorithm guarantees that, if an interpretation that is a bisimulation exists, then it will enumerate at least one of them. The algorithm iterates through each possible assignment of each implementation *reaction* to be interpreted as a given formal reaction or trivial; for each assignment, iterates through each partial specification of an interpretation that satisfies the reactions assigned to be formal; then sets up the remaining trivial reactions as a system of equations and finds a minimal solution. By testing each enumerated interpretation with either of the permissive condition tests described above, this algorithm will find an interpretation or assert that none exists.

Theorem 6. *Whether a bisimulation interpretation exists from a given implementation CRN to a given formal CRN is PSPACE-complete.*

Proof (Sketch). The above algorithm runs in polynomial space, thus proving membership. The reactions that simulate a Turing machine in Theorem 5 restrict the interpretation enough that any interpretation other than the one given (up to a permutation of the formal species) will be invalid.   □

Theorem 7. *When the number of reactants in a formal reaction k is bounded by a constant, whether a bisimulation interpretation exists is NP-complete.*

Proof (Sketch). If a valid interpretation exists, the above algorithm guarantees that it will produce a valid polynomial-size interpretation which can be checked in polynomial time by Theorem 4. A 3-SAT formula with clauses e.g. $(x_1 \lor \neg x_2 \lor x_3)$ can be encoded in implementation reactions e.g. $s_C \rightarrow x_1^t + x_2^f + x_3^t$. Additional reactions $s_T \rightleftharpoons x_i^t + x_i^f$ restrict interpretations that satisfy the delimiting condition to correspond to satisfying assignments of the 3-SAT formula.   □

5   Discussion

Comparing Chemical Reaction Networks on different levels of abstraction is an important tool for systematic programming with CRNs. We showed how to adapt the concept of bisimulation to check whether one CRN is a correct implementation of another. We showed that bisimulation can be used to prove the correctness of some existing CRN implementations, and to identify subtle but real problems with others. We discussed transitivity and modularity, which can be used to simplify a bisimulation proof. We presented different algorithms to check bisimulation which are adapted to different cases. We showed that the condition can be checked in polynomial time with favorable assumptions, is NP-complete with less favorable assumptions, and is PSPACE-complete in the general case.

Algorithms such as the graph-updating algorithm and loopsearch algorithm scale better with the number of meaningful species than the number of null species, while engineered CRN implementations generally do not use loops that produce null species. Thus those algorithms will be faster than their worst-case limits in practical cases. For example, the graph-updating algorithm takes at most $(2zn^k + 1)n^k = O(n^{2k+1})$ cycles in theory, where n is the number of implementation species, k the largest number of reactants in a formal reaction, and z the number of implementation species with empty interpretation. When there are no null species (or when none can be produced in a loop, as in schemes such as [16]), this becomes at most n^k cycles.

In CRN bisimulation, we require that *every* implementation species has an interpretation as a (possibly empty) multiset of formal species. In contrast, verification methods such as pathway decomposition [15] or serializability [9] both assume that each formal species is represented by one implementation species, while other implementation species are classified into fuels, wastes, and intermediates. Because of this, pathway decomposition and serializability compare formal reactions to implementation pathways which begin and end with (representations of) formal species, while in bisimulation an individual implementation reaction can be interpreted and compared to the formal CRN. An additional consequence, for pathway decomposition, is that correctness guarantees do not apply to implementation states that cannot be reached from initial states representing formal species, whereas bisimulation is more robust in that correctness is asserted in those cases as well. Furthermore, even in the permissive condition, bisimulation requires that there *exist* an implementation pathway which implements a given formal reaction, while pathway decomposition and serializability both require that *all* implementation pathways have properties which may be nontrivial to check. This locality is what allows us to prove the complexity results given, which we suspect are significantly lower complexity than methods that depend on implementation pathways.

However, the use of interpretations instead of pathways means that in some cases CRN bisimulation and pathway decomposition differ on which implementations they consider correct. Bisimulation can easily be adapted to situations where there is no clear single "canonical representation" of a given formal species, while pathway decomposition has difficulty. For example, the implementation in [12] of the *reversible* formal reaction $A+B \rightleftharpoons C+D$ by reversible implementation reactions $\{x_A \rightleftharpoons i_A, i_A + x_B \rightleftharpoons i_{CD}, i_{CD} \rightleftharpoons x_C + i_D, i_D \rightleftharpoons x_D\}$. Bisimulation considers this correct with the obvious interpretation, while pathway decomposition considers the ability to release x_C then reverse without releasing x_D to be an error. On the other hand, bisimulation has trouble handling implementation species with no well-defined interpretation. Shin describes a "delayed choice" phenomenon where an implementation CRN commits to implementing one of two formal reactions before deciding which one, producing an intermediate that cannot be correctly interpreted as either of the reaction's products or their reactants; such implementations are generally considered incorrect according to bisimulation but pathway decomposition often considers them correct [15]. Shin proposes a hybrid notion of correctness where an implementation CRN is considered correct if it is a correct implementation according to pathway decomposition of some intermediate CRN, and the intermediate CRN is a correct implementation of the formal CRN according to bisimulation [15]. This notion considers correct any implementation that is correct according to either pathway decomposition or bisimulation, plus some others.

One area this theory overlooks is the rates of reactions and the probabilities of reaching certain states. For example, in [16] Soloveichik et al. argue that the concentration of each intermediate is proportional to the product of that of the formal species which we would call its interpretation, and thus the reaction rates are approximately correct. Whether this can be generalized, and whether bisimulation can help this generalization, is an important open question.

Acknowledgements. The authors would like to thank Chris Thachuk, Damien Woods, Dave Doty, and Seung Woo Shin for helpful discussions. RFJ and EW were supported by NSF grants 1317694, 1213127, and 0832824. RFJ was supported by Caltech's Summer Undergraduate Research Fellowship program and an NSF graduate fellowship. QD's current affiliation is Epic Systems, Madison, Wisconsin.

References

1. Angluin, D., Aspnes, J., Eisenstat, D.: A simple population protocol for fast robust approximate majority. Distrib. Comput. **21**, 87–102 (2008)
2. Cardelli, L.: Two-domain DNA strand displacement. Math. Struct. Comput. Sci. **23**, 247–271 (2013)
3. Cardelli, L.: Morphisms of reaction networks that couple structure to function. BMC Syst. Biol. **8**, 1–18 (2014)
4. Cardelli, L., Csikász-Nagy, A.: The cell cycle switch computes approximate majority. Sci. Rep. **2** (2012)
5. Chen, Y.J., Dalchau, N., Srinivas, N., Phillips, A., Cardelli, L., Soloveichik, D., Seelig, G.: Programmable chemical controllers made from DNA. Nat. Nanotechnol. **8**, 755–762 (2013)
6. Dong, Q.: A bisimulation approach to verification of molecular implementations of formal chemical reaction networks. Master's thesis, Stony Brook University (2012)
7. Garey, M.R., Johnson, D.S.: Computers and Intractability: A Guide to the Theory of NP-Completeness. W. H. Freeman & Co., New York (1979)
8. Grun, C., Sarma, K., Wolfe, B., Shin, S.W., Winfree, E.: A domain-level DNA strand displacement reaction enumerator allowing arbitrary non-pseudoknotted secondary structures. CoRR (2015). http://arxiv.org/abs/1505.03738
9. Lakin, M.R., Stefanovic, D., Phillips, A.: Modular verification of chemical reaction network encodings via serializability analysis. Theor. Comput. Sci. **632**, 21–42 (2016)
10. Lakin, M.R., Youssef, S., Polo, F., Emmott, S., Phillips, A.: Visual DSD: a design and analysis tool for DNA strand displacement systems. Bioinformatics **27**, 3211–3213 (2011)
11. Milner, R.: Communication and Concurrency. Prentice-Hall Inc., Upper Saddle River (1989)
12. Qian, L., Soloveichik, D., Winfree, E.: Efficient turing-universal computation with DNA polymers. In: Sakakibara, Y., Mi, Y. (eds.) DNA 16 2010. LNCS, vol. 6518, pp. 123–140. Springer, Heidelberg (2011)
13. Rackoff, C.: The covering and boundedness problems for vector addition systems. Theor. Comput. Sci. **6**(2), 223–231 (1978)
14. Savitch, W.J.: Relationships between nondeterministic and deterministic tape complexities. J. Comput. Syst. Sci. **4**, 177–192 (1970)
15. Shin, S.W., Thachuk, C., Winfree, E.: Verifying chemical reaction network implementations: a pathway decomposition approach. CoRR (2014). http://arxiv.org/abs/1411.0782
16. Soloveichik, D., Seelig, G., Winfree, E.: DNA as a universal substrate for chemical kinetics. Proc. Nat. Acad. Sci. **107**, 5393–5398 (2010)
17. Srinivas, N.: Programming chemical kinetics: engineering dynamic reaction networks with DNA strand displacement. Ph.D. thesis, California Institute of Technology (2015)

A Coarse-Grained Model of DNA Nanotube Population Growth

Vahid Mardanlou[1], Leopold N. Green[2], Hari K.K. Subramanian[3],
Rizal F. Hariadi[4], Jongmin Kim[4], and Elisa Franco[3(✉)]

[1] Electrical and Computer Engineering, University of California at Riverside,
Riverside, CA 92521, USA
`vahid.mardanlou@email.ucr.edu`
[2] Bioengineering, University of California at Riverside, Riverside, CA 92521, USA
`leopold.green@email.ucr.edu`
[3] Mechanical Engineering, University of California at Riverside,
Riverside, CA 92521, USA
`{harikks,efranco}@engr.ucr.edu`
[4] Wyss Institute for Biologically Inspired Engineering, Harvard University,
Boston, MA 02115, USA
`{rizal.hariadi,jongmin.kim}@wyss.harvard.edu`

Abstract. We derive a coarse-grained model that captures the temporal evolution of DNA nanotube length distribution during growth experiments. The model takes into account nucleation, polymerization, joining, and fragmentation processes in the nanotube population. The continuous length distribution is segmented, and the behavior of nanotubes in each length bin is modeled using ordinary differential equations. The binning choice determines the level of coarse graining. This model can handle time varying concentration of tiles, and we foresee that it will be useful to model dynamic behaviors in other types of biomolecular polymers.

Keywords: DNA nanotubes · Ordinary differential equations · Growth · Dynamic DNA nanotechnology

1 Introduction

Many biological scaffolds, such as the cytoskeleton, are built with filamentous polymers that are constantly assembling and disassembling in response to environmental inputs and cellular instructions. DNA nanotechnology has produced a variety of artificial, rationally designed tubular structures whose dimensions and mechanical properties are comparable to those of natural filaments such as actin and microtubules [1–5]. DNA nanotubes self-assemble from tiles that can be single-stranded or multi-stranded; inter-tile binding patterns are determined by programmable single stranded interaction domains. Here, we focus on DNA nanotubes assembling from multi-stranded tiles. We present a continuous time, coarse-grained approach to model how the nanotube length distribution varies over time in a population of nanotubes. This research is motivated by the rapid

© Springer International Publishing Switzerland 2016
Y. Rondelez and D. Woods (Eds.): DNA 2016, LNCS 9818, pp. 135–147, 2016.
DOI: 10.1007/978-3-319-43994-5_9

expansion of dynamic DNA nanotechnology, which offers exciting opportunities to use DNA strand displacement circuits to control tile self-assembly [6]. Existing methods to model tile or nanotube assembly are not suited to track length distributions, and cannot handle dynamically varying free tile concentration.

DNA tile assembly has been primarily modeled with algorithmic approaches, which explore the capacity of various tile interaction rules to produce desired patterns, and investigate their computational power [7–9]. The famous Winfree kTAM model models the kinetics of individual tiles binding and unbinding to a growing structure [7]; single molecule [10] and single-filament [11] measurements have been used to validate this model. Nanotube length distributions can be computed using kTAM stochastic simulations [12,13]; nucleation and end-joining events can be taken into account, however this approach is only viable at low (nanomolar) tile concentrations. The most notable limitation of this model is the fact that it cannot handle time-varying concentration of free tiles.

We formulate and validate a phenomenological model for self-assembling nanotubes that describes how the length varies over time in a population of nanotubes. This model is coarse-grained in the sense that the nanotube population is segmented by length in a number of bins, and we use ordinary differential equations to describe how the population of each bin varies over time. We model several processes that are known to affect nanotube length: nucleation, polymerization and depolymerization, end-joining, and fragmentation.

The model is fitted to growth data of DAE-E tile nanotubes described by Rothemund et al. where the total tile concentration is constant. However, our model can easily handle scenarios where the total tile concentration varies over time (for example, by activation or deactivation [6]), because it is naturally compatible with ODE chemical reaction networks. These scenarios will be considered in future work. We expect that this model will be useful to describe other self-assembling polymer systems operating at different time and length scales.

2   Results

2.1   Model Derivation

We consider a solution including assembled nanotubes and unpolymerized tiles. The real distribution of nanotubes is continuous, because our sample includes tubes having any length $l \in [0, l_{max}]$, where l_{max} is the maximum observed length (or a physically meaningful upper bound for length). To build a model that is computationally tractable, we segment the population of molecular species present in the system. We assume the sample includes tiles, whose concentration is indicated as T; nucleated assemblies of tiles, or nuclei, whose concentration is L_0; nanotubes, which are binned by length, so that variable L_n indicates the concentration of nanotubes in bin n. The bin width, which we indicate as l_b, can be chosen depending on the acceptable level of coarseness (and complexity) of the model, because it determines the number of species. For example, if l_b is 300 nm, variable L_1 is the concentration of tubes of length 300 nm. If $l_{max} = 30\,\mu m$, the number of variables in the model is $n_{max} = \lfloor l_{max}/l_b \rfloor = 100$. Segmentation introduces implicitly the assumption that a tube can switch from bin n to bin $n \pm 1$ only if it acquires or loses a number n_b of tiles, which are the tiles forming a tube segment of length l_b. As an example, let us take again $l_b = 300$ nm; let us assume that the nanotube circumference is on average 7 tiles, each ≈ 14 nm wide; then we find $n_b = 147$ (these figures are based on previous measurements on DAE-E tile nanotubes [1], and were confirmed in our experiments).

We finally assume that tiles, nuclei, and tubes form and interact via several processes, which cause changes in the segmented distribution of nanotube length. An overview of these processes is provided below, together with phenomenological expressions for the rates at which these processes occur. For each description, we identify an equivalent, phenomenological reaction that describes how tiles, nuclei, and nanotubes interact in our model.

Nucleation. Tile assembly is a cooperative process: there is a minimum number of tiles that need to bind simultaneously to form a nucleus, from which polymerization of a nanotube can be initiated. We assume that nucleation depends on the concentration of tiles, and proceeds with rate $k_{nucl} T^{n_{nucl}}$, where n_{nucl} is the critical nucleation size. The equivalent phenomenological reaction describing nucleation is:

$$n_{nucl} T \xrightarrow{\ k_{nucl}\ } L_0. \tag{1}$$

Polymerization and Depolymerization. Nuclei and nanotubes grow as tiles bind to accessible sites. The polymerization rate depends on the concentration of tiles as well as the availability of binding sites: for tubes of length n, polymerization occurs at rate $k_p T L_n$. For nuclei, which are smaller patches of tiles, we hypothesize a different polymerization rate $k_{p_0} T L_0$. Tiles can also dissociate from tubes (and nuclei) at a rate that depends exclusively on the concentration of tubes: for tubes of length n, the depolymerization rate is $k_d L_n$; for nuclei, we consider a different depolymerization rate $k_{d_0} L_0$. Equivalent phenomenological reactions describing polymerization and depolymerization are:

$$n_b T + L_n \xrightarrow{\ k_p\ } L_{n+1}, \quad n \geq 1 \tag{2}$$

$$(n_b - n_{nucl}) T + L_0 \xrightarrow{\ k_{p_0}\ } L_1 \tag{3}$$

$$L_1 \xrightarrow{\ k_d\ } L_0 + (n_b - n_{nucl}) T \tag{4}$$

$$L_{n+1} \xrightarrow{\ k_d\ } L_n + n_b T , \text{ for } n \geq 1 \tag{5}$$

$$L_0 \xrightarrow{\ k_{d_0}\ } n_{nucl} T \tag{6}$$

Here, the stoichiometric coefficients indicate how many tiles need to be added or removed from a nanotube in a certain bin length so that it moves to an adjacent bin. These coefficients are however not related to the order of the reaction rate; for example, in reaction (2) it is not required that n_b tiles bind simultaneously to the tube, therefore this is a second order reaction.

End-Joining. Nucleated nanotubes diffusing in solution grow not only by polymerization, but also by end-joining, as demonstrated experimentally in [14].

We assume that the joining rate depends on the length of the nanotubes, on their diameter d, and on the concentration of nanotubes in the corresponding length bins. For example, if we consider length bins n and m, we postulate that the joining rate of nanotubes in these bins is $k_{join}(n, m)L_n L_m$. An estimate for $k_{join}(n, m)$ is given in expression (7), which is derived in detail in Ref. [15]. This expression assumes that DNA nanotubes are rigid rods, and that their end-joining is a diffusion controlled reaction:

$$k_{join}(m, n) = \frac{\alpha}{l_b}\left[\frac{1}{m}\ln\left(\frac{ml_b}{d}\right) + \frac{1}{n}\ln\left(\frac{nl_b}{d}\right)\right] \tag{7}$$

where $\alpha = (12\kappa k_B \mathcal{T} d)/\eta$. Here η is the dynamic viscosity of the liquid, k_B is the Boltzmann constant, $\mathcal{T}$ is the absolute temperature, d is the nanotube diameter, and κ is a factor accounting for the fraction of productive nanotube collisions. Note that each joining reaction can occur by joining of either end of each nanotube, so every reaction should be accounted for twice.

The concentration of nanotubes in a given bin n increases when shorter tube end-join; an example equivalent reaction is:

$$L_{n-m} + L_m \xrightarrow{k_{join}(n - m, m)} L_n \quad \text{for} \quad 1 \leq m \leq \min\{n_{max} - n, n - 1\}.$$

We observe that order of the reactants in the above reaction does not matter. For example, consider the bin of nanotubes having length $5l_b$. The end-joining reactions that contribute to an increase in the concentration L_5 are:

$$L_1 + L_4 \xrightarrow{k_{join}(1, 4)} L_5$$

$$L_2 + L_3 \xrightarrow{k_{join}(2, 3)} L_5$$

Reactions $L_4 + L_1 \xrightarrow{k_{join}(1, 4)} L_5$ and $L_3 + L_2 \xrightarrow{k_{join}(2, 3)} L_5$ are redundant and should not be included in the mass balance. The concentration of tubes in bin of length nl_b is therefore incremented by only $\lfloor \frac{n}{2} \rfloor$ end-joining reactions, where "$\lfloor \ \rfloor$" is the largest integer less than or equal to $\frac{n}{2}$.

The concentration of nanotubes in bin n also decreases due to end-joining events, as exemplified in this reaction:

$$L_n + L_m \xrightarrow{k_{join}(m, n)} L_{n+m} \quad \text{for} \quad 1 \leq m \leq n_{max} - n$$

Fragmentation. Formed nanotubes can spontaneously break into shorter fragments. We assume that breakage rate of nanotubes of length $n\,l_b$ depends on the concentration of nanotubes in bin length n via a constant rate k_{break}. For simplicity, we also assume fragments can only form to fall into the given length bins, and that a nanotube can break into at most two fragments per unit time. For example, a nanotube of length $n \, l_b$ can break into two fragments of length $m l_b$ and $(n-m) l_b$. The equivalent phenomenological reaction for fragmentation is:

$$L_n \xrightarrow{k_{break}} L_m + L_{n-m}.$$

The rate at which the concentration of nanotubes in bin n decreases due to fragmentation is $(n-1) k_{break} L_n$, because there are $n-1$ sites at which breakage may occur. For example, consider the concentration of tubes having length $5 l_b$. In this bin, fragmentation occurs according to the following equivalent reactions:

$$L_5 \xrightarrow{k_{break}} L_1 + L_4, \quad L_5 \xrightarrow{k_{break}} L_2 + L_3,$$
$$L_5 \xrightarrow{k_{break}} L_3 + L_2, \quad L_5 \xrightarrow{k_{break}} L_4 + L_1.$$

We note that the concentration of nanotubes in bin n also increases due to breakage of longer nanotubes (bins $m \geq n+1$).

Tile Activation and Deactivation. Tiles may be injected or removed from the system, for example by activation or deactivation reactions that could be performed via strand displacement. Expression to model these phenomena depend on the mechanism chosen for activation and deactivation; at this stage we assume they are some continuous functions $a(t)$ (activation) and $d(t)$ (deactivation).

We now combine the phenomena described above and their rates to write a set of differential equation describing the evolution of our segmented nanotube length distribution. Equations (8)–(11) describe the time derivative of the concentration of tiles (T), nuclei (L_0) and tubes in bin n (L_n):

$$\frac{dT}{dt} = a(t) - d(t) - \underbrace{n_{nucl} k_{nucl} T^{n_{nucl}}}_{\text{nucleation}} - \underbrace{n_b k_p T \sum_{i=1}^{n_{max}-1} L_i - (n_b - n_{nucl}) k_{p_0} T L_0}_{\text{polymerization}}$$

$$+ \underbrace{n_b k_d \sum_{i=2}^{n_{max}} L_i + (n_b - n_{nucl}) k_d L_1 + n_{nucl} k_{d_0} L_0}_{\text{depolymerization}}, \tag{8}$$

$$\frac{dL_0}{dt} = \underbrace{k_{nucl} T^{n_{nucl}}}_{\text{nucleation}} - \underbrace{k_{p_0} T L_0}_{\text{polymerization}} + \underbrace{k_d L_1 - k_{d_0} L_0}_{\text{depolymerization}}, \tag{9}$$

$$\frac{dL_1}{dt} = \underbrace{k_{p_0} T L_0 - k_p T L_1}_{\text{polymerization}} + \underbrace{k_d L_2 - k_d L_1}_{\text{depolymerization}} - \underbrace{2 \sum_{m=1}^{n_{max}-1} k_{join}(1,m) L_1 L_m - 2 k_{join}(1,1) L_1^2}_{\text{end-joining}}$$

$$+ \underbrace{2 \sum_{m=2}^{n_{max}} k_{break}(L_m \to L_1 + L_{m-1}) L_m}_{\text{fragmentation}}, \tag{10}$$

$$\underbrace{\frac{dL_n}{dt} = k_p T(L_{n-1} - L_n|_{n<n_{max}})}_{\text{polymerization}} + \underbrace{k_d(L_{n+1}|_{n<n_{max}} - L_n)}_{\text{depolymerization}}$$

$$\underbrace{- 2 \sum_{m=1}^{n_{max}-n} k_{join}(n,m)L_n L_m - 2k_{join}(n,n)L_n^2|_{n\leq\frac{n_{max}}{2}} + 2 \sum_{m=1}^{\lfloor\frac{n}{2}\rfloor} k_{join}(n-m,m)L_{n-m}L_m}_{\text{end-joining}}$$

$$\underbrace{+ 2 \sum_{m\geq n+1}^{n_{max}} k_{break}(L_m \rightarrow L_n + L_{m-n})L_m - (n-1)k_{break}(L_n \rightarrow L_m + L_{n-m})L_n,}_{\text{fragmentation}}$$

$$n = 2, ..., n_{max}. \tag{11}$$

We clarify the presence of end-joining term $-2k_{join}(n,n)L_n^2$ in Eq. (11). This term is needed for two reasons: first, end-joining can occur at either end of a nanotube; second, when two nanotubes of length n end-join, two of them are lost from bin n. Thus, overall Eq. (11) should include a term $-4k_{join}(n,n)L_n^2$: we chose to split it between sum $-2\sum_{m=1}^{n_{max}-n} k_{join}(n,m)L_n L_m$ and the isolated term $-2k_{join}(n,n)L_n^2$.

In this model, the mass transfer (species concentration conversion) is carefully balanced. If $a(t) = d(t) = 0$, i.e. there is no net injection or removal of tiles, the total concentration of tiles T^{tot} remains constant. Equation (12) describes tile mass conservation:

$$T^{tot} = T + n_{nucl}L_0 + n_b \sum_{n=1}^{n_{max}} nL_n, \quad \text{when} \quad a(t) = d(t) = 0. \tag{12}$$

Our model naturally guarantees tile mass conservation, because differential equations (8)–(11) satisfy the following equality:

$$\frac{dT}{dt} + n_{nucl}\frac{dL_0}{dt} + n_b \sum_{n=1}^{n_{max}} n\frac{dL_n}{dt} = 0, \quad \text{where} \quad a(t) = d(t) = 0.$$

2.2 Modeling Nanotube Growth

We test the capacity of model (8)–(11) to fit nanotube length distributions in a sample of growing nanotubes where there is no injection or removal of tiles, i.e. $a(t) = d(t) = 0$.

Experiments. We consider the DAE-E tile nanotubes described in [1,14] (Fig. 1A); these tiles assemble from five distinct DNA strands, and are about 14.3 nm wide. Tiles form nanotubes that generally have a 7-tile circumference, and a diameter $d = 13 \times 10^{-3}$ (μm). Strand 3 of each tile is labeled with Cy3, so that nanotube length can be measured in fluorescence microscopy experiments [1,14]. Sequences and protocols are reported in the Methods section.

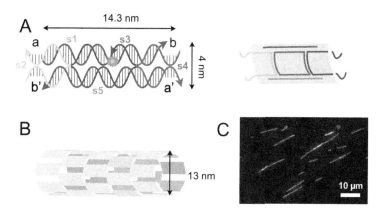

Fig. 1. A: Schematic of the DAE-E tile used in our experiments (left) and abstraction (right); tiles interact via their complementary sticky end domains a, a', b and b'. Strand 3 (s3) is labeled with a Cy3 fluorophore. B: Tiles interact to form tubes via their complementary sticky ends [1]; C: Example fluorescence microscopy image of nanotubes.

Tile annealing was stopped at the expected melting temperature of sticky ends, 47 °C, so that the nuclei and short tube population is initially negligible [14] (see Methods section for details). We immediately started imaging the sample; no nanotubes or short assemblies are initially visible in fluorescence microscopy images right after quenching. Samples were incubated at room temperature, and imaged using a fluorescence microscope during the following 30 h to assess changes in the nanotube length distribution over time. This experiment was run in triplicate samples; Fig. 2D shows mean length and standard deviation of the mean measured across experiments. We observe rapid growth in the first five hours, where the mean nanotube length reaches 5 μm; the mean length slowly reaches 10 μm in the following 25 h. The measured growth curve in our experiments is qualitatively similar to previous experiments by Ekani-Nkodo et al. [14]; however our samples achieve on average a 40 % higher mean length, presumably because of higher tile concentration.

Data Fitting. We fitted Model (8)–(11) to the nanotube distributions that were experimentally measured over time. First, we established parameters of our model that depend on the type of nanotubes we consider, and on the desired level of coarse graining. We choose the length bin: $l_b = 300$ nm; the maximum nanotube length is set as $l_{max} = 33$ μm. These choices imply that we have a number of bins $n_{max} = 110$; thus the model includes 112 differential equations (including the ODEs for tiles and nuclei). The number of tiles in a tube chunk of length l_b is $n_b = 147$. (This follows from our assumption that the tubes have a 7-tile circumference, with ≈14 nm tile width). ODEs were integrated with in-house MATLAB scripts. We chose an integration step of 10 s.

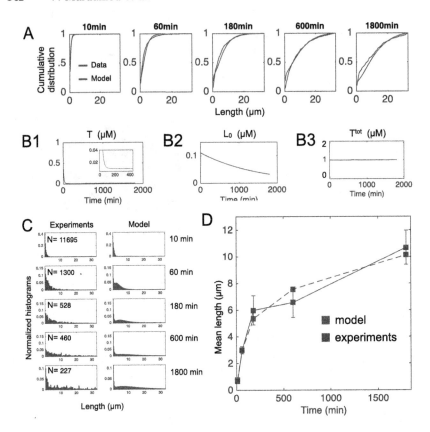

Fig. 2. Performance of our coarse-grained model. A: Comparison between a representative experimental (blue) cumulative length distributions gathered during nanotube growth and the cumulative distribution generated by our model (red). B1 Time course of free tile concentration. B2: Nuclei concentration. B3: Total tile concentration remains constant. C: Histograms of representative growth experiment (blue) and histograms generated by our model (red). N indicates the number of nanotubes measured in each sample (given equal measurement surface). D: Mean and standard deviation of the mean in our nanotube growth experiments, compared with the mean length predicted by our fitted model. (Color figure online)

Predicted nanotube distributions at time t are generated deterministically by integrating ODEs (8)–(11) (up to time t). Rather than comparing length histograms generated by the model to experimental histograms, we compare their cumulative distributions. This choice is motivated by the following observations: (1) Cumulative distributions are by definition scaled by the sample size (in our case, nanotube sample number). This implies that cumulative distributions of different samples can be immediately compared without requiring *ad hoc* normalization (which would be necessary to compare histograms). (2) Cumulative distributions do not require binning of data like a histogram, thus the fitting procedure is not biased by a choice of bin width.

To compare experimental and simulated cumulative distributions at the same points (nanotube length), we interpolated experimental cumulative length distributions. After interpolation, the model and experimental distributions at any given time are described by comparable vectors:

$$V_{sim}(t) = \frac{1}{\mathcal{L}_{sim}} \cdot \begin{bmatrix} L_{1,sim}(t) \\ L_{1,sim}(t) + L_{2,sim}(t) \\ \vdots \\ \sum_1^{n_{max}} L_{i,sim}(t) \end{bmatrix}, \quad V_{exp}(t) = \frac{1}{\mathcal{L}_{exp}} \cdot \begin{bmatrix} L_{1,exp}(t) \\ L_{1,exp}(t) + L_{2,exp}(t) \\ \vdots \\ \sum_1^{n_{max}} L_{i,exp}(t) \end{bmatrix},$$

where $\mathcal{L}_{sim} = \sum_{i=1}^{n_{max}} L_{i,sim}(t)$, and $\mathcal{L}_{exp} = \sum_{i=1}^{n_{max}} L_{i,exp}(t)$. These distributions are to be compared at times t = 10, 30, 60, 180, 600, 1800 min.

Fitting is done to identify several parameters in our model. Specifically, we fit the nucleation rate (k_{nucl}) and critical nucleation size (n_{nucl}), the polymerization and depolymerization rates for tubes and nuclei (k_p, k_{po}, k_d, and k_{do}), the breakage rate (k_{break}), and parameter α in the joining rate expression (7). These parameters can be stacked in a vector p, and we set up our fitting problem as the minimization of the objective function, which simultaneously compares the simulated distribution to the distributions of three separate experiments:

$$\min_p J = \sum_{j=1}^{3} \sum_t (V_{sim}(t) - V_{exp}^j(t))^\top (V_{sim}(t) - V_{exp}^j(t)). \tag{13}$$

Minimization was done using the `fmincon` routine in MATLAB. Initial conditions for the parameters were sampled uniformly in physically plausible intervals delimited by lower and upper bounds listed in Table 1; parameters were also constrained to fall within these bounds. We also imposed a lower bound of 10 nM on the admissible free tile concentration. Initial conditions for the model variables were chosen as $T(0) = 1 \mu M$, $L_i(0) = 0 \mu M$ for $i = 0, ..., n_{max}$, i.e. we assumed that initially only free tiles are present; this assumption is consistent with the experimental conditions that were fitted, because nanotube annealing was quenched quickly from 47 °C to room temperature so the concentration of nuclei and tubes is negligible. This approximation is sensible, however it will be refined in the future by measuring the initial concentration of tiles, nuclei and short tubes using native gel electrophoresis or fluorescence spectroscopy.

Several fitting campaigns were launched and evaluated. Minimizing the objective function (13) is a non-convex problem, thus the fitting routine is likely to converge to local minima that depend on the randomly chosen initial conditions for the parameter vector. We report one of the best fitting results in Table 1.

Figure 2A shows the performance of the model in fitting cumulative distributions in one of our experiments. The concentration of free tiles and nuclei as a function of time are shown in Fig. 2B1 and B2; during the time course, the total concentration of tiles remains constant as shown in Fig. 2B3 (this is a sanity check that mass conservation equation (12) holds).

Table 1. Fitting parameters and results. This table lists the parameters of Model (8)–(11) that were fitted, including lower bound (L.B.) and upper bounds (U.B.) used in the fitting procedure.

Parameter	Units	L.B.	U.B.	Fitted value	Definition
k_p	$M^{-1}s^{-1}$	10^1	10^7	4.699×10^3	Tube polymerization rate
k_d	s^{-1}	10^{-8}	10^4	1.192×10^{-4}	Tube depolymerization rate
α	$\mu m\ M^{-1}s^{-1}$	10^{-4}	10^{11}	1.030×10^5	End-joining parameter
k_{break}	s^{-1}	10^{-10}	10^2	1.255×10^{-6}	Fragmentation rate
k_{nucl}	$M^{1-n_{nucl}}s^{-1}$	10^{-10}	10^4	5.746×10^3	Nucleation rate
k_{p_0}	$M^{-1}s^{-1}$	10^1	10^7	5.261×10^1	Nuclei polymerization rate
k_{d_0}	s^{-1}	10^{-8}	10^1	1.096×10^{-5}	Nuclei depolymerization rate
n_{nucl}	N/A	2	5	4.999	Critical nucleation size

We also tested the capacity of our model to generate histograms that reflect the measurements; experimental data were binned consistently to our model bin width $l_b = 300$ nm; results are shown in Fig. 2C. We also computed the mean nanotube length from simulated distributions, and we compared it with the measured mean length in Fig. 2D (standard deviation is computed over triplicate experiments).

3   Discussion and Conclusion

We derived a deterministic coarse-grained model to describe the growth kinetics of DNA nanotubes. The model takes into account nucleation, polymerization and depolymerization, joining and fragmentation processes. We obtained macroscopic rates for each of these processes by fitting our model to experimental cumulative distributions of nanotube length at different time points. Future work will focus on (a) assessing the capacity of the model to fit different experimental conditions (for example temperature, tile concentration, and ionic conditions), and (b) validating the model against non-fitted scenarios. Although the model presented here is not yet validated, we briefly discuss our fitting results in relation to the literature. Rate constants are listed in Table 1; our fitted polymerization rate of tiles to growing tubes is about two orders of magnitude slower than rates estimated for individual tiles binding to a growing tube [10,11]; however, if we factor into our fitted rate the number of tiles in a bin, we obtain a rate directly comparable to the literature [10,11]. The fitted tube depolymerization rate is about two orders of magnitude slower than the estimates in [11] for DAO-O nanotubes measured at 33 °C. Our combined nucleation rate $n_{nucl}\,k_{nucl} \approx 2.87 \times 10^4\ M^{1-n_{nucl}}/s$ is within one order of magnitude of previous estimates in [6]; our critical nucleation size is twice that estimated in [6]. Polymerization and depolymerization rates on growing nuclei are significantly slower than the corresponding rates of growing tubes; while this expectation is sensible for polymerization (due to a reduced number of binding sites), depolymerization rates should be consistent with those of formed nanotubes. Our joining rate expression (7) depends on the fitted parameter α, and on the length of joining tubes: k_{join} peaks when $n = m = 1$, reaching 2.15×10^6/M/s, and is a decreasing function of nanotube length. We evaluated the expression for k_{join} over its domain n, m, and computed an average $k_{join} = 2.63 \times 10^5$/M/s. This rate is one order of magnitude higher than the joining rate estimate for DNA ribbons in [12], which is 3.5×10^4/M/s. Finally, our fragmentation rate is in the order of 10^{-6}/M/s, suggesting that breakage events are extremely rare. It should be noted that fitted rates depend on the chosen level of granularity; further work and validation are required to identify tradeoffs between model granularity and its predictive capacity.

The complexity of DNA nanotube formation and growth justifies the formulation of a model based on length distribution segmentation. Unlike most polymers or copolymers that form linear chains where each link of the chain is a single chemical compound [16], DNA nanotubes are hollow cylinders where several monomers participate in each growth layer. This feature makes DNA nanotubes similar to microtubules [17], and makes kinetic modeling of assembly significantly more complex.

Our modeling approach relies on ordinary differential equations that are typically used in polymerization process control models [16], however to track length distributions while maintaining our model numerically tractable we define chemical species that represent tube populations in a certain length bin.

Analytical solutions can be found for simple filament models that capture nucleation, growth and breakage by using linearization and moment methods [18], but this approach may not be applicable to our model. Conformational dynamics of DNA filaments are outside the scope of this research [19]. Future work will focus on modeling nanotube length distributions where tiles are activated or deactivated over time [6].

4   Methods

DNA Sequences. All DNA was purchased from Integrated DNA Technologies (Coralville, IA). Sequences were taken from Rothemund et al. [1]. Following the notation introduced in Fig. 1, sequences are reported below.

s1: 5'-CTCAGTGGACAGCCGTTCTGGAGCGTTGGACGAAACT,

s2: 5'-GTCTGGTAGGCACCACTGAGAGGTA,

s3: 5'-T-Cy3-CCAGAACGGCTGTGGCTAAACAGTAACCGAAGCACCAACGCT,

s4: 5'-CAGACAGTTTCGTGGTCATCGTACCT,

s5: 5'-CGATGACCTGCTTCGGTTACTGTTTAGCCTGCTCTAC.

Sample Preparation. Lyophilized DNA oligonucleotides were resuspended in water, quantitated by UV absorbance at 260 nm using Thermo Scientific Nanodrop 2000c Spectrophotometer, and stored at $-20\,°C$. DAE-E Tile tube solution was prepared to target $1\,\mu M$ tile concentration by adding $1\,\mu M$ (final concentration) of s1, s2, s3, s4, and s5 strands of DNA, 1x Tris-Acetate-EDTA (TAE) diluted buffer and 12.4 mM MgCl$_2$. After vortexing for 60 s, tube solution was placed in Eppendorf Mastercycler PCR machine and annealed by heating solution to 90 °C, and cooled to 47 °C over a 6 h period. Annealing was stopped at 47 °C, and samples were rapidly quenched to room temperature.

Image Acquisition and Processing. Samples were imaged on an inverted microscope (Nikon Eclipse TI-E) with 60X/1.40 NA oil immersion objectives. Aliquots of nanotube solution (diluted to 50 nM tile concentration) were deposited on Fisherbrand microscope cover glass 12-545E No. 1 (thickness = 0.13 to 0.17 mm; Size: 50 × 22 mm). VWR Micro Slides (Plain, Selected, Pre-cleaned, 25 × 75 mm, 1.0 mm thick) were placed gently on the cover glass. Nanotubes were imaged using Cy3 filter cube (Semrock Brightline - Cy3-404C-NTE-ZERO). Upon image capturing, exposure time was set to 90 ms. ImageJ plugin AnalizeSkeleton was used to obtain nanotube contours, from which we estimated tube length using a MATLAB script. All branched nanotubes are automatically eliminated from our sample. Due to camera limitations, tubes below $0.3\,\mu m$ in length were discarded from tube length distributions.

Acknowledgement. The authors thank Deborah K. Fygenson, Bernard Yurke, Rebecca Schulman, and Martha Grover for advice and discussions. This research was supported by DE grant SC0010595.

References

1. Rothemund, P.W.K., Ekani-Nkodo, A., Papadakis, N., Kumar, A., Fygenson, D.K., Winfree, E.: Design and characterization of programmable DNA nanotubes. J. Am. Chem. Soc. **126**(50), 16344–16352 (2004)
2. Liu, D., Park, S.H., Reif, J.H., LaBean, T.H.: DNA nanotubes self-assembled from triple-crossover tiles as templates for conductive nanowires. Proc. Natl. Acad. Sci. U.S.A. **101**(3), 717–722 (2004)
3. Mitchell, J.C., Harris, J.R., Malo, J., Bath, J., Turberfield, A.J.: Self-assembly of chiral DNA nanotubes. J. Am. Chem. Soc. **126**(50), 16342–16343 (2004)
4. Sharma, J., Chhabra, R., Cheng, A., Brownell, J., Liu, Y., Yan, H.: Control of self-assembly of DNA tubules through integration of gold nanoparticles. Science **323**(5910), 112–116 (2009)
5. Douglas, S.M., Chou, J.J., Shih, W.M.: DNA-nanotube-induced alignment of membrane proteins for NMR structure determination. Proc. Natl. Acad. Sci. **104**(16), 6644–6648 (2007)
6. Zhang, D.Y., Hariadi, R.F., Choi, H.M.T., Winfree, E.: Integrating DNA strand-displacement circuitry with DNA tile self-assembly. Nat. Commun. **4** (2013)
7. Winfree, E.: Simulations of computing by self-assembly. Technical report, California Institute of Technology (1998)
8. Reif, J.H., Sahu, S., Yin, P.: Compact error-resilient computational DNA tilings. In: Chen, J., Jonoska, N., Rozenberg, G. (eds.) Nanotechnology: Science and Computation. Natural Computing Series, pp. 79–103. Springer, Heidelberg (2006)

9. Soloveichik, D., Winfree, E.: Complexity of self-assembled shapes. SIAM J. Comput. **36**(6), 1544–1569 (2007)
10. Evans, C.G., Hariadi, R.F., Winfree, E.: Direct atomic force microscopy observation of DNA tile crystal growth at the single-molecule level. J. Am. Chem. Soc. **134**(25), 10485–10492 (2012)
11. Hariadi, R.F., Yurke, B., Winfree, E.: Thermodynamics and kinetics of DNA nanotube polymerization from single-filament measurements. Chem. Sci. **6**(4), 2252–2267 (2015)
12. Schulman, R., Winfree, E.: Synthesis of crystals with a programmable kinetic barrier to nucleation. Proc. Natl. Acad. Sci. **104**(39), 15236–15241 (2007)
13. Mohammed, A.M., Schulman, R.: Directing self-assembly of DNA nanotubes using programmable seeds. Nano Lett. **13**(9), 4006–4013 (2013)
14. Ekani-Nkodo, A., Kumar, A., Fygenson, D.K.: Joining and scission in the self-assembly of nanotubes from DNA tiles. Phys. Rev. Lett. **93**(26), 268301 (2004)
15. Hariadi, R.F., Winfree, E., Yurke, B.: Determining hydrodynamic forces in bursting bubbles using DNA nanotube mechanics. Proc. Natl. Acad. Sci. **112**(45), E6086–E6095 (2015)
16. Fogler, H.S.: Elements of Chemical Reaction Engineering, 4th edn. Prentice-Hall International London, Upper Saddle River (2005)
17. Flyvbjerg, H., Jobs, E., Leibler, S.: Kinetics of self-assembling microtubules: an "inverse problem" in biochemistry. Proc. Natl. Acad. Sci. **93**(12), 5975–5979 (1996)
18. Knowles, T.P., Waudby, C.A., Devlin, G.L., Cohen, S.I., Aguzzi, A., Vendruscolo, M., Terentjev, E.M., Welland, M.E., Dobson, C.M.: An analytical solution to the kinetics of breakable filament assembly. Science **326**(5959), 1533–1537 (2009)
19. Andrews, S.S.: Methods for modeling cytoskeletal and DNA filaments. Phys. Biol. **11**(1), 011001 (2014)

On the Runtime of Universal Coating for Programmable Matter

Zahra Derakhshandeh[1]([⊠]), Robert Gmyr[2], Alexandra Porter[1],
Andréa W. Richa[1], Christian Scheideler[2], and Thim Strothmann[2]

[1] Computer Science, CIDSE, Arizona State University, Tempe, USA
{zderakhs,amporte6,aricha}@asu.edu
[2] Department of Computer Science, Paderborn University, Paderborn, Germany
{gmyr,scheidel,thim}@mail.upb.de

Abstract. Imagine coating buildings and bridges with smart particles (also coined smart paint) that monitor structural integrity and sense and report on traffic and wind loads, leading to technology that could do such inspection jobs faster and cheaper and increase safety at the same time. In this paper, we study the problem of uniformly coating objects of arbitrary shape in the context of *self-organizing programmable matter*, i.e., the programmable matter consists of simple computational elements called particles that can establish and release bonds and can actively move in a self-organized way. Particles are anonymous, have constant-size memory and utilize only local interactions in order to coat an object. We continue the study of our Universal Coating algorithm by focusing on its runtime analysis, showing that our algorithm terminates within a *linear number of rounds* with high probability. We also present a matching linear lower bound that holds with high probability. We use this lower bound to show a *linear lower bound on the competitive gap* between fully local coating algorithms and coating algorithms that rely on global information, which implies that our algorithm is also optimal in a competitive sense. Simulation results show that the competitive ratio of our algorithm may be better than linear in practice.

1 Introduction

Inspection of bridges, tunnels, wind turbines, and other large civil engineering structures for defects is a time-consuming, costly, and potentially dangerous task. In the future, *smart coating* technology, or *smart paint*, could do the job more efficiently and without putting people in danger. The idea behind smart coating is to form a thin layer of a specific substance on an object which then makes it possible to measure a condition of the surface (such as temperature or cracks) at any location, without direct access to the location. The concept of smart coating

Z. Derakhshandeh, A. Porter and A.W. Richa—Supported in part by NSF grants CCF-1353089, CCF-1422603, and REU–026935.

R. Gmyr, C. Scheideler and T. Strothmann—Supported in part by DFG grant SCHE 1592/3-1.

Y. Rondelez and D. Woods (Eds.): DNA 2016, LNCS 9818, pp. 148–164, 2016.
DOI: 10.1007/978-3-319-43994-5_10

already occurs in nature, such as proteins closing wounds, antibodies surrounding bacteria, or ants surrounding food to transport it to their nest. These diverse examples suggest a broad range of applications of smart coating technology in the future, including repairing cracks or monitoring tension on bridges, repairing space craft, fixing leaks in a nuclear reactor, or stopping internal bleeding. We continue the study of coating problems in the context of self-organizing programmable matter consisting of simple computational elements, called particles, that can establish and release bonds and can actively move in a self-organized way using the geometric version of the amoebot model presented in [1,2]. In doing so, we proceed to investigate the runtime analysis of our Universal Coating algorithm, introduced in [3]. We first show that coating problems do not only have a (trivial) linear lower bound on the runtime, but that there is also a linear lower bound on the competitive gap between the runtime of fully local coating algorithms and coating algorithms that rely on global information. We then investigate the worst-case time complexity of our Universal Coating algorithm and show that it terminates within a linear number of rounds with high probability (w.h.p.)[1], which implies that our algorithm is optimal in terms of worst-case runtime and also in a competitive sense. Moreover, our simulation results show that in practice the competitive ratio of our algorithm is often better than linear.

1.1 Amoebot Model

In the *geometric amoebot model*, we consider the graph G_{eqt}, where $G_{eqt} = (V_{eqt}, E_{eqt})$ is the infinite regular triangular grid graph. Each vertex in V_{eqt} is a position that can be occupied by at most one particle (see Part (a) of Fig. 1). Each particle occupies either a single node or a pair of adjacent nodes in G_{eqt}. Any structure a particle system can form can be represented as a subgraph of G_{eqt}. Two particles occupying adjacent nodes are *connected* by a *bond*, and we refer to such particles as *neighbors*. The bonds do not only ensure that the particles form a connected structure but they are also used for exchanging information as explained below.

Particles move by executing a series of *expansions* and *contractions*. If a particle occupies one node it is *contracted* and can expand to an unoccupied adjacent node to occupy two nodes. If it occupies two nodes it is *expanded* and can contract to occupy a single node. In part(b) of Fig. 1, we illustrate a set of particles on the underlying graph G_{eqt}. For an expanded particle, we denote the node the particle last expanded into as the *head* of the particle and the other occupied node is called its *tail*. For a contracted particle, the single node occupied by the particle is both its head and its tail.

To stay connected as they move, neighboring particles coordinate their motion in a *handover*, which can occur in two ways. A contracted particle p can

---

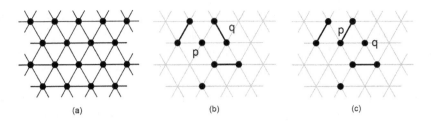

Fig. 1. (a) shows a section of G_{eqt}, where nodes of G_{eqt} are shown as black circles. (b) shows five particles on G_{eqt}; the underlying graph G_{eqt} is depicted as a gray mesh; a contracted particle is depicted as a single black circle and an expanded particle is depicted as two black circles connected by an edge. (c) depicts the resulting configuration after a handover was performed by particles p and q in (b).

initiate a handover by expanding into a node occupied by an expanded neighbor q. Thus, p "pushes" q and forces it to contract. Alternatively, an expanded particle q can initiate a handover by contracting. While contracting, q "pulls" a neighboring contracted particle p to the node it is vacating, thereby forcing p to expand. Parts (b) and (c) of Fig. 1 illustrate two particles labeled as p and q performing a handover. Particles are *anonymous* but each particle has a collection of uniquely labeled *ports* corresponding to the edges incident to the nodes the particle occupies. Bonds between adjacent particles are formed through ports that face each other. The particles are assumed to have a common *chirality*, meaning they all have the same notion of *clockwise (CW) direction*. This allows each particle p to label its ports counting in clockwise direction; without loss of generality, we assume each particle labels its head and tail ports from 0 to 5. However particles can have different offsets of the labelings, so they do not share a common sense of orientation. Each particle hosts a local memory of constant size for which any neighboring particle has read and write access. With this mechanism particles can communicate by writing into each others memory. The *configuration* C of the system at the beginning of time t consists of the nodes in G_{eqt} occupied by the object and the set of particles, and for each particle p, C contains the current state of p, including whether it is expanded or contracted, its port labeling, and the contents of its local memory.

Following the standard asynchronous model of computation [4], we assume that the system progresses through atomic *activations* of individual particles. At each (atomic) activation a particle can perform at most one movement and an arbitrary bounded amount of computation, involving its local memory and the shared memories of its neighbors. A classical result under this model is that for any asynchronous execution of atomic particle activations, we can organize these activations sequentially still producing the same end configuration [4]. We count (asynchronous) time in terms of the number of activations. A *round* is over once each particle has been activated at least once. We assume the activation sequence to be *fair*, i.e., for each particle p and any point in time t, p will eventually be activated at some time $t' \geq t$.

1.2   Universal Coating Problem

In the *universal coating problem* we consider an instance (P, O) where P represents the particle system and O represents the fixed object to be coated. Let $V(P)$ be the set of nodes occupied by P and $V(O)$ be the set of nodes occupied by O (when clear from the context, we may omit the $V(\cdot)$ notation). Then let the set of nodes in G_{eqt} neighboring O be called the *surface (coating) layer*. Let n be the number of particles and B_1 be the number of nodes in the surface layer. For any two nodes $v, w \in V_{eqt}$, the *distance* $d(v, w)$ between v and w is the length of the shortest path in G_{eqt} from v to w. The distance $d(v, U)$ between a $v \in V_{eqt}$ and $U \subseteq V_{eqt}$ is defined as $\min_{w \in U} d(v, w)$. An instance is *valid* if the following properties hold:

1. The particles are all contracted and are initially in an *idle* state.
2. The subgraphs of G_{eqt} induced by $V(O)$ and $V(P) \cup V(O)$, respectively, are connected, i.e., there is a single object and the particle system is connected to the object.
3. The subgraph of G_{eqt} induced by $V_{eqt} \setminus V(O)$ is connected, i.e., the object O has no holes.[2]
4. $V_{eqt} \setminus V(O)$ is $2(\lceil \frac{n}{B_1} \rceil + 1)$-connected, i.e. O cannot form *tunnels* of width less than $2(\lceil \frac{n}{B_1} \rceil + 1)$.

Note that a width of at least $2\lceil \frac{n}{B_1} \rceil$ is needed to guarantee that the object can be evenly coated. The coating of narrow tunnels requires specific technical mechanisms that complicate the protocol without contributing to the basic idea of coating, so we ignore such cases in favor of simplicity.

A configuration C is *legal* if and only if all particles are contracted and

$$\min_{v \in V_{eqt} \setminus (V(P) \cup V(O))} d(v, V(O)) \geq \max_{v \in V(P)} d(v, V(O))$$

meaning that all particles are as close to the object as possible or *coat O as evenly as possible*. If the object has to be coated by more than one layer of particles then the *i-th layer* around the object are the nodes that have a distance of i to the object.

An algorithm *solves* the universal coating problem if, starting from any valid configuration, it reaches a *stable legal configuration* C in a finite number of rounds. A configuration C is said to be stable if no particle in C ever performs a state change or movement.

1.3   Related Work

Many approaches have been proposed with potential applications in smart coating; these can be categorized as active and passive systems. In passive systems particles move only based on their structural properties and interaction with

---

[2] If O does contain holes, we consider the subset of particles in each connected region of $V_{eqt} \setminus V(O)$ separately.

the environment, or have only limited computational ability and lack control of their motion. Examples include DNA self-assembly systems (see, e.g., the surveys in [5–7]), population protocols [8], and slime molds [9,10]. Our focus is on active systems, in which computational particles control their actions and motions to complete specific tasks. Coating has been extensively studied in the area of *swarm robotics*. However, coating of objects is commonly not studied as a stand-alone problem, but is part of *collective transport* (e.g., [11]) or *collective perception* (e.g., see respective section of [12]). Some research focuses on coating objects as an independent task under the name of *target surrounding* or *boundary coverage*. The techniques used in this context include stochastic robot behaviors [13,14], rule-based control mechanisms [15] and potential field-based approaches [16]. While the analytic techniques developed in swarm robotics are somewhat relevant to this work, those systems have more computational power and movement capabilities as those studied in this work. Michail and Spirakis recently proposed a model [17] for network construction inspired by population protocols [8]. The population protocol model is related to self-organizing particle systems but is different in that agents (corresponding to our particles) can move freely in space and establish connections at any time. It would, however, be possible to adapt their approach to study coating problems under the population protocol model. In [3] we presented our Universal Coating algorithm and proved its correctness. We also showed it to be worst-case work-optimal, where work is measured in terms of number of particle movements.

1.4   Our Contributions

In this paper we continue the analysis of the *Universal Coating algorithm* introduced in [3]. As our main contribution in this paper, we investigate the runtime of our algorithm and prove that our algorithm terminates within a *linear number of rounds* with high probability. We also present a matching linear lower bound for local-control coating algorithms that holds with high probability. We use this lower bound to show a *linear lower bound on the competitive gap* between fully local coating algorithms and coating algorithms that rely on global information, which implies that our algorithm is also optimal in a competitive sense. We then present some simulation results demonstrating that in practice the competitive ratio of our algorithm is often much better than linear.

Overview. In Sect. 2, we present a brief overview of the algorithm presented in [3]. We present a comprehensive formal runtime analysis of our algorithm, by first presenting some lower bounds on the competitive ratio of any local-control algorithm in Sect. 3, and then proving that our algorithm has a runtime of $O(n)$ rounds w.h.p. in Sect. 4, which matches our lower bounds. Due to space limitations, full proofs were omitted and can be found in [18].

2   Universal Coating Algorithm

In this section we summarize the Universal Coating algorithm introduced in [3]. This algorithm is constructed by combining a number of asynchronous primitives, which are integrated seamlessly without any underlying synchronization. The *spanning forest* primitive organizes the particles into a spanning forest, which determines the movement of particles while preserving system connectivity; the *complaint-based coating* primitive controls the coating of the first layer, by expanding the coating of the first layer while there is still room and there are still particles not yet touching the object; the *general layering* primitive allows each layer i to form only after layer $i - 1$ has been completed, for $i \geq 2$; and a *node-based leader election* primitive elects a position (in B_1) whose occupant becomes a leader particle, which is used to trigger the layering process after the first layer. Due to space constraints we will only briefly describe the basic ideas of the Universal Coating algorithm (see [3] for a detailed description). For completeness the pseudocode for all algorithmic primitives can be found in [18].

2.1   Preliminaries

We define the set of *states* that a particle can be in as *idle*, *follower*, *root*, and *retired*. In addition to its state, a particle maintains a constant number of other flags, which in our context are constant size pieces of information visible to neighboring particles. A flag x owned by some particle p is denoted by $p.x$. Recall that a *layer* is the set of nodes v in G_{eqt} that are equidistant to the object O. A particle keeps track of its current layer number in $p.layer$. In order to respect the constant-size memory constraint of particles, we take all layer numbers modulo 4. We say that layer i, $i \geq 1$, is *complete* if each node in that layer is occupied with a retired particle (except for the last layer which can be partially filled with retired particles). Each root particle p has a flag storing a port label $p.down$, which points to an occupied node of the object if $p.layer$ (the layer p occupies) equals 1, and to an occupied node adjacent to its head in layer $p.layer - 1$ if $p.layer > 1$. Next we describe the coating primitives in more detail.

2.2   Coating Primitives

The **spanning forest primitive** organizes the particles in a spanning forest $\mathcal{F}$, which creates a straightforward mechanism for particles to move while preserving connectivity (see [1,19] for details). Initially, all particles are *idle*. A particle p touching the object changes its state to *root*. For any other idle particle p we use the rule that once p sees a root or a follower in its neighborhood, it stores the direction to one of them in $p.parent$, changes its state to *follower*, and generates a complaint flag. Follower particles use handovers to follow their parents and update $p.parent$ as they move so that it always points to the same parent (resp. the follower that took over the role of p's parent q because of a handover with q). In this way, the trees formed by the parent relations stay connected, only use positions they have covered before, and do not mix with other trees.

Roots p use the flag $p.dir$ to determine their movement direction. As a root p moves, it updates $p.dir$ so that it always points to the next position of a clockwise movement around the object. For any particle p, we call the particle occupying the position that $p.parent$ resp. $p.dir$ points to the *predecessor* of p. If a root particle does not have a predecessor, we call it a *super-root*.

The **complaint-based coating primitive** is used for the coating of the first layer. Each time a particle p holding at least one complaint flag is activated, it forwards one to its predecessor as long as that predecessor does not hold more than two complaint flags. Allowing each particle to hold up to two complaint flags has two reasons: it ensures that a constant size memory is sufficient for storing the complaint flags and that the flags quickly move forward to the super-roots. A contracted super-root p only expands to $p.dir$ if it holds at least one complaint flag, and when it expands, it consumes one of these complaint flags. All other roots p move towards $p.dir$ whenever possible (i.e., no complaint flags are needed for that) by performing a handover with a predecessor (which must be another root) or a successor (which is a root or a follower of its tree), with preference given to a follower so that one more particle reaches layer 1. As we will see, these rules ensure that whenever there are particles in the system that are not yet at layer 1, eventually one of these particles will move to layer 1, unless layer 1 is already completely filled with contracted particles.

The **leader election primitive** runs during the complaint-based coating primitive to elect a position along layer 1 as the leader position. This algorithm is similar to the algorithm presented in [1] with the difference that leader candidates are associated with positions instead of particles (which is important because in our case particles may still move during the leader election primitive) as we presented in [3]. The primitive only terminates once all positions in layer 1 are occupied. Once the leader position is determined, and there are no more followers or all positions in layer 1 are filled by contracted particles, then whatever particle currently covers that position becomes the *leader*. If the primitive does not terminate (which only happens if $n < B_1$ and layer 1 is never completely covered), then the complaint flags ensure that the super-roots eventually stop, which eventually results in a stable legal coating. Given that a particle becomes the leader, that leader becomes a marker particle that marks a neighboring position at the next higher layer as a *marked position* and changes to the *retired* state. The marked position determines the point at which root particles should align in the next higher layer. Once a contracted root p has a retired particle in the direction $p.dir$, it retires as well, which causes the particles in layer 1 to change to the *retired* state in a counter-clockwise order. Also, the general layering primitive becomes active, which builds subsequent layers until there are no longer followers in the system.

In the **general layering primitive**, whenever a follower is connected to a retired particle, it changes to the root state. Root particles continue to move along positions of their layer in a clockwise (if the layer number is odd) or counter-clockwise (if the layer number is even) direction till they reach the marked position in that layer, or a retired particle in that layer, or a previously empty position of a lower layer (which causes them to change direction). Complaint flags are no longer needed to move to empty spots. Followers follow their parents as before. A contracted root particle may retire due to the following reasons: (i) it is located at the marked position and the marker particle in the lower layer tells it that all particles in that layer are already retired (which it can determine locally), or (ii) it has a retired particle in the direction of its layer. Once a particle at a marked position retires, it becomes a marker particle and marks a neighboring position at the next higher layer as a marked position.

3   Lower Bounds

Recall that a *round* is over once every particle in P has been activated at least once. The *runtime* $T_A(P, O)$ of a coating algorithm A is defined as the worst-case number of rounds (over all sequences of particle activations) it takes for A to solve the coating problem (P, O). Certainly, there are instances (P, O) where every coating algorithm has a runtime of $\Omega(|P|)$ (see Lemma 1), though there are also many other instances where the coating problem can be solved much faster. Since a worst-case runtime of $\Omega(|P|)$ is fairly large and therefore not very helpful to distinguish between different coating algorithms, we intend to study the runtime of coating algorithms relative to the best possible runtime.

A coating algorithm A is called *c-competitive* if for any valid instance (P, O),

$$\mathrm{E}[T_A(P, O)] \leq c \cdot \mathrm{OPT}(P, O) + C$$

where $\mathrm{OPT}(P, O)$ is the minimum runtime needed to solve the coating problem (P, O) and C is a value independent of (P, O). Unfortunately, also for the competitiveness a high lower bound holds for all local-control algorithms.

Lemma 1. *The worst-case runtime required by any local-control algorithm to solve the universal coating problem is $\Omega(|P|)$ with high probability.*

Theorem 1. *Any local-control algorithm that solves the universal coating problem has a competitive ratio of $\Omega(|P|)$.*

Therefore, even the competitive ratio can be very high in the worst case. We will revisit the notion of competitiveness in Sect. 5.

4   Worst-Case Number of Rounds

In this section we show that our algorithm solves the coating problem within a linear number of rounds w.h.p. We start with some basic notation in Sect. 4.1. Section 4.2 presents a useful tool that allows us to consider initial configurations that consists of a forest of paths, and Sect. 4.3 presents a simpler synchronous parallel model for particle activations that we can use to analyze the worst-case number of rounds. Section 4.4 presents the analysis of the number of rounds required to build the first coating layer. Finally, we analyze the number of rounds required to complete all other coating layers, once layer 1 has been completed (Sect. 4.5).

4.1   Preliminaries

We start with some notation. Let B_i denote the number of nodes of G_{eqt} at distance i from object O (i.e., B_i denotes the number of nodes on layer i), and let M be the highest index of a final coating layer for n particles, i.e., M is such that $\sum_{j=1}^{M-1} B_j < n \leq \sum_{j=1}^{M} B_j$. Layer i is *complete* if it is completely filled with contracted retired particles, for $i < M$, or, if $i = M$, all particles have reached their final position, are contracted and never move again. A *legal coating* of i layers for object O consists of a configuration where the first i layers are complete. Let n_i denote the number of particles of the system that will not belong to layers 1 through $i - 1$, i.e. $n_i = n - \sum_{j=1}^{i-1} B_j$.

Given a configuration C, we define a directed graph $A(C)$ over all nodes in G_{eqt} occupied by *active* (i.e., either follower or root) particles in C. For every expanded active particle in C, $A(C)$ contains a directed edge from the tail to the head node of p. For every follower p, $A(C)$ has a directed edge from the head of p to $p.parent$, and for every root particle p, p has a directed edge from its head to the node in the direction of $p.dir$, if $p.dir$ is occupied by an active particle. Certainly, since every node has at most one outgoing edge in $A(C)$, the nodes of $A(C)$ can only form a collection of disjoint trees or a ring of trees. Note that a ring of trees may emerge in any layer, but the leader election primitive ensures that this is only temporarily the case in layer 1, because once a leader or marker is elected, it retires, causing the ring in $A(C)$ to break. In the runtime analysis for layer i, $i \geq 2$, we always assume that layers 1 to $i - 1$ are already complete, which means that any particle reaching the marker position on layer i stops and retires (hence, we never have to consider a ring of trees for higher layers). The super-roots defined in Sect. 2 correspond to the roots of the trees in $A(C)$. For any subset T of $A(C)$, let $|T|$ denote the number of particles located at T. We may abuse notation at times, when clear from the context, and refer to the set of particles located at T simply as T.

4.2   Debranching

In this section, we present a useful tool that allows us to show that for any tree of particles T in the spanning forest $\mathcal{F}$ right at the start of the construction of layer i, the worst-case configuration of T with respect to the number of asynchronous rounds required to construct layer i is when T consists of a single line path. Let $C^{(A)}(i)$ denote the first configuration when layer i becomes complete according to the execution of our asynchronous algorithm, $1 \leq i \leq M$. Consider an asynchronous particle activation sequence α starting from configuration $C^{(A)}(i - 1)$ until we reach $C^{(A)}(i)$. Let $Layer(i)$ denote the worst-case number of asynchronous rounds for any such α, $2 \leq i \leq M$.

Lemma 2 (Debranching). *Let T be an arbitrary tree in a spanning forest $\mathcal{F}$ at $C^{(A)}(i - 1)$. One can always construct a tree T' consisting of a single path containing the same set of particles as T, such that for any (asynchronous) activation sequence α starting from $\mathcal{F}$, we build an (asynchronous) activation* *sequence α' starting from $\mathcal{F} \backslash \{T\} \cup \{T'\}$ such that both α and α' only reach a legal coating for i layers at the end of the sequence, and the number of asynchronous rounds in α is no more than that of α'.*

Repeated applications of Lemma 2 for each of trees in $\mathcal{F}$ at $C^{(A)}(i-1)$ allows us to bound the worst case for $Layer(i)$ starting from a set of disjoint line paths $L_1, \ldots, L_x$ that each contain a single root particle in layer i. Note that some of these paths may be connected through other root particles in layer i (i.e., when we look at $A(C^{(A)}(i-1))$ we may see a collection of trees composed by the union of some of the L_j paths and root nodes on layer i).

4.3   From Asynchronous to Parallel Schedules

In this section, we show that instead of analyzing our algorithm for asynchronous activations of particles, it suffices to consider a much simpler model of parallel activations of particles. A *movement schedule* is a sequence of particle system configurations $C_0, C_1, \ldots, C_t$.

A movement schedule is called a (valid) *parallel schedule* if every C_j represents a valid configuration of a connected particle system (i.e., each particle is either expanded or contracted, and every node of G_{eqt} is occupied by at most one particle) and for every $j \geq 0$, C_{j+1} is reached from C_j in a way that for every particle p one of the following properties holds:

1. p occupies the same positions in C_j and C_{j+1}, or
2. p expands into an adjacent node that was empty in C_j, or
3. p contracts, leaving the node occupied by its tail empty in C_{j+1}, or
4. p is part of a handover with a neighboring particle p'.

This means that several particles are allowed to move from C_j to C_{j+1}, but at most one contraction or expansion per particle is possible.

A *tree schedule* $\mathcal{S} = (L, (C_0, \ldots, C_t))$ is a valid movement schedule $C_0, \ldots, C_t$ with the property that $A(C_0)$ is a tree T_0, L is a simple path in G_{eqt} along some nodes $v_1, \ldots, v_\ell$ with v_1 being occupied by the head of the super-root in $A(T_0)$, and for the other nodes $v \in L$, $v \notin T_0$. Note that the root of a tree T in $A(C)$, for some configuration C, is occupied by the unique super-root particle located in T. We also require that in $C_0, \ldots, C_t$ all particles have to follow the unique path in $A(T_0) \cup L$ from their initial position in C_0 to v_ℓ in a way that $A(C_j)$ is a single tree for every $j \geq 1$ (which implies that $A(C_j)$ covers a subset of $T_0 \cup L$). A tree schedule is *greedy* if all particles move in a greedy manner, i.e., they perform a contraction, expansion, or handover in the direction of their unique path whenever possible as long as this does not violate the constraints of a tree schedule. For any configuration C of the tree schedule and any involved particle p we define its *head distance* (resp. *tail distance*) $d_h(p, C)$ (resp. $d_t(p, C)$) to be the number of edges along the unique path of p from the head (resp. tail) of p to v_ℓ. Certainly, $d_h(p, C) \in \{d_t(p, C), d_t(p, C) - 1\}$, depending on whether p is contracted or expanded. For any two configurations C_1 and C_2 and any particle

p we say that C_1 *dominates* C_2 w.r.t. p (or short, $C_1(p) \geq C_2(p)$) if and only if $d_h(p, C_1) \leq d_h(p, C_2)$ and $d_t(p, C_1) \leq d_t(p, C_2)$. Altogether, we say that C_1 *dominates* C_2 if and only if C_1 dominates C_2 with respect to every particle p.

Consider now an arbitrary fair asynchronous activation sequence of particles in the given particle tree, and let $C_j^{(A)}$, $0 \leq j \leq t$, be the configuration of the particles at the end of asynchronous round j if the particles move as prescribed by our coating algorithm. Then it holds:

Lemma 3. *For any particle tree extended by a root path L, and any fair asynchronous activation sequence, there is a greedy tree schedule $(L, (C_1, C_2, \ldots))$ so that $C_j^{(A)} \geq C_j$ for all $j \geq 1$.*

This lemma also holds for *forest schedules*, i.e., we have multiple trees with roots at different places of some root path which are following that path in the same direction. Hence, once we have an upper bound for the time it takes for a forest schedule to reach a final configuration, we also have an upper bound for the number of rounds it takes for any fair asynchronous activation sequence to reach the final configuration.

We can extend the dominance argument to also hold for complaint flags, if we follow the rules of the complaint-based coating primitive for the forwarding and consumption of these flags: Basically, besides enforcing that the dominance should hold for the d_h and d_t functions, we also show that dominance holds with respect to the distance function $d_c(f, C)$, which gives the number of edges along the unique path from the node containing the particle that currently holds flag f to v_ℓ. We also make the parallel schedules more restrictive, by allowing each particle to hold at most *one* (rather than two, as described in the asynchronous algorithm) complaint flag. In the context of parallel schedules, a particle p holding a complaint flag f in C_j can forward f to an adjacent particle q, if q did not hold any complaint flag in C_j (and is not receiving a complaint flag from any other neighbor at round C_j), or q holds a complaint flag f' in C_j which will be forwarded in parallel to one of q's neighbors in this round. Hence, we obtain the following lemma:

Lemma 4. *For any particle forest grounded at a root path L, and any fair asynchronous activation sequence, there is a forest schedule $(L, (C_1, C_2, \ldots))$ so that with or without the use of complaint flags, $C_j^{(A)} \geq C_j$ for all $j \geq 1$.*

Line Schedules. We will look at a particular case of parallel tree schedules, called *(parallel) line schedules*, where $A(C_0)$ is equal to a simple path (basically $L \cup A(C_0)$ induces a line of particles in G_{eqt}). Line schedules are a "simpler" variant of tree schedules and will be very useful in computing an upper bound on the worst-case number of asynchronous rounds for building a layer i, $i \geq 2$, in Sect. 4.5. Here, we prove some general helper lemmas about greedy line schedules. We start with the following lemma.

Lemma 5. *Consider any greedy line schedule* $\mathcal{S} = (L, (C_0, C_1, \ldots, C_t))$. *Let* L' *be the path given by* $L \cup A(C_0)$. *If we start from a configuration* C_0 *with no two adjacent expanded particles along* L', *then for all* j, $0 \leq j \leq t$, C_j *never has two adjacent expanded particles in* L'.

This implies that the super-root is able to move forward every two steps (as whenever it is expanded, the particle behind it must be contracted), implying the following lemma.

Lemma 6. *Consider any greedy line schedule* $\mathcal{S} = (L, (C_0, C_1, \ldots, C_t))$ *with* m *particles and* $L = (v_1, \ldots, v_\ell)$ *and let* $L' = L \cup A(C_0)$. *If we start from a configuration* C_0 *with no two adjacent expanded particles along* L', *then it takes at most* $2(\ell + k)$ *configurations for any* $k \leq m$ *until* $v_{\ell-k+1}, \ldots, v_\ell$ *are occupied by contracted particles.*

Lemmas 4 and 6 tremendously simplify the analysis of the runtime to complete layer 1 or a higher layer respectively, as we will see.

4.4   First Layer: Complaint-Based Coating and Leader Election

At the beginning of our algorithm, the spanning forest needs to be formed. It is easy to bound the runtime for that:

Lemma 7. *Following the spanning forest primitive, the particles form a spanning forest within* $O(n)$ *rounds.*

By Lemma 2, we can upper bound on the worst-case number of asynchronous rounds by assuming that $\mathcal{F}$ consists of a forest of lines initially. Every particle that joins the spanning forest as a follower generates a complaint flag, which needs to be forwarded along the parent connections to the root of its $A(C)$-tree (which we called a super-root) so that it moves forward. In the following assume that the super-roots just move along a common path L (and not a ring, which they really do).

For ease of presentation, we assume that $B_1 \leq n$ (the proofs can be easily extended to also cover the case $B_1 > n$ [18]); we also assume that the root of a tree also generates a complaint flag upon its activation (this assumption does not hurt our argument since it only increases the number of the flags generated in the system). Let $\mathcal{S} = (L, (C_0, C_1, \ldots, C_t))$ be a greedy forest schedule where C_0 is the first configuration after the spanning forest has been formed, C_t is the first configuration in $\mathcal{S}$ when layer 1 becomes complete and L is the set of nodes on layer 1.

The following lemma shows that the complaint flags will be consumed in $\mathcal{S}$ as soon as they reach a super-root, or if that is not possible, that the algorithm still makes steady progress by either bringing more complaint flags onto layer 1 or by moving all the remaining complaint flags in the system one position closer to a super-root within two consecutive parallel rounds. Note that it becomes easier to prove Lemma 8 since we know that $\mathcal{F}$ in the context of $A(C_j)$ consists of a forest of lines.

Lemma 8. *Consider a round C_j of $\mathcal{S}$, where $0 \leq j \leq t-2$. Then within the next two parallel rounds of $\mathcal{S}$ (i.e., in configurations C_{j+1} or C_{j+2}), we must have that (i) at least one complaint flag is consumed, (ii) at least one more complaint flag reaches a particle in layer 1, (iii) all the remaining complaint flags move one position closer to a super-root along L, or (iv) layer 1 is completely filled (possibly with some expanded particles).*

We use Lemma 8 to show first that layer 1 will be filled with particles (some possibly still expanded) in $O(n)$ rounds. From that point on, in another $O(n)$ rounds, one can guarantee that expanded particles in layer 1 will each contract in a handover with a follower particle, and hence all particles in layer 1 will be contracted, resulting in the following lemma:

Lemma 9. *After $O(n)$ rounds, layer 1 must be filled with contracted particles.*

Once layer 1 is filled, the leader election primitive can proceed. Our leader election result [1] implies the following runtime bound.

Lemma 10. *Within $O(n)$ further rounds, a position of layer 1 has been elected as the leader position, w.h.p.*

Once a leader position has been elected and there is either no more followers (if $n \leq B_1$) or all positions are completely filled by contracted particles (which can be checked in an additional $O(B_1)$ rounds), the particle currently occupying the leader position is elected as the leader. Once a leader has been found, the particles on layer 1 retire, which takes $O(B_1)$ further rounds. Putting it all together, we get:

Corollary 1. *The worst-case number of rounds for the Universal Coating algorithm to complete layer 1 is $O(n)$, w.h.p.*

4.5   Higher Layers

We will now use the results we have proven for line schedules in Sect. 4.3 in order to prove an upper bound on the worst-case number of rounds — denoted by $Layer(i)$ — for building layer i once layer $i-1$ is complete, for $2 \leq i \leq M$. As we have seen in Sect. 4.2, we can assume that the spanning forest $\mathcal{F}$ at $C^{(A)}(i-1)$ consists of a forest of lines, when bounding $Layer(i)$.

There are two type of movements of root particles that occur when coating a layer i: (a) root particles that move through the nodes in layer i in the set direction dir (CW or CCW) for layer i, and (b) root particles that move through the nodes in layer $i+1$ in the opposite direction, over the already retired particles in layer i. We bound the worst-case scenario for each of the two types of movement independently, in order to get a an upper bound on $Layer(i)$. If we want to maximize the number of movements (also inducing the worst-case number of rounds) in case (a) for a forest of lines in $\mathcal{F}$ at $C^{(A)}(i-1)$, one would have just a single line L_a of n_i particles whose root is as far from the marker particle as possible according to direction dir for layer i (since this would maximize the distances that the particles would need to traverse until they retired). Similarly for case (b) we would have a single line L_b of particles as further from the marker particle as possible according to the opposite direction of dir. Let $L = v_1, \ldots, v_{B_i}$ be the nodes on layer i listed in the order that they appear in layer i from the marker position v_1 following direction dir. From Lemma 3, we can upper bound the number of rounds required to complete all movements of type case (a) using the line schedule $\mathcal{S} = (L, (C_0, \ldots, C_t))$, where $A(C_0) = L_a$. It then follows from Lemma 6 that it would take $O(B_i)$ asynchronous rounds for the process in case (a) to terminate. A similar argument can be used to show that all the movements in case (b) terminate in $O(B_{i+1}) = O(B_i)$ rounds (note that, trivially, $B_{i+1} \leq 6B_i$). This implies the following lemma:

Lemma 11. *Starting from configuration $C^{(A)}(i-1)$, the worst-case additional number of asynchronous rounds for layer i to also become complete is $O(B_i)$.*

Putting it altogether, for layers 2 through M:

Corollary 2. *The worst-case number of rounds for the Universal Coating algorithm to coat layers 2 through M is $O(\sum_{i=2}^{M} B_i) = O(n)$.*

Combining Corollaries 1 and 2 we get, for any given valid object O to be coated, and any initial valid configuration of the set of particles P:

Theorem 2. *The total number of asynchronous rounds required for our algorithm to reach a legal coating configuration, starting from an arbitrary valid initial configuration (P, O) is $O(n)$ w.h.p., where n is the number of particles in the system.*

5   Simulation Results

In this section we present a brief simulation based analysis of our algorithm which shows that in practice our algorithm exhibits a better than linear average competitive ratio. Since $\text{OPT}(P, O)$ (as defined in Sect. 3) is in general difficult to compute, we investigate the competitiveness with the help of an appropriate lower bound for $\text{OPT}(P, O)$. Recall the definitions of the distances $d(p, q)$ and $d(p, U)$ for $p, q \in V_{\text{eqt}}$ and $U \subseteq V_{\text{eqt}}$. Consider any valid instance (P, O). Let $\mathcal{L}$ be the set of all legal particle positions of (P, O), that is, $\mathcal{L}$ contains all sets $U \subseteq V_{\text{eqt}}$ such that the positions in U constitute a coating of the object O by the particles in the system.

We compute a lower bound on $\text{OPT}(P, O)$ as follows. Let $\mathcal{L}$ be defined as above and let $U \in \mathcal{L}$. Consider the complete bipartite graph $G(P, U)$ with P and U being the partitions of the graph. For each edge $e = (p, q) \in P \times U$ set the cost of the edge to $w(e) = d(p, q)$. Every perfect matching in $G(P, U)$ corresponds to an assignment of the particles to positions in the coating given by U. The maximum edge weight in a matching corresponds to the maximum distance a

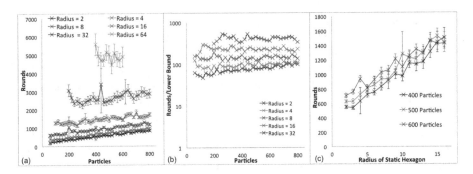

Fig. 2. (a) Number of rounds varying number of particles   (b) Ratio of number of rounds to lower bound (log scale)   (c) Number of rounds varying static hexagon radius

particle has to travel in order to coat the object accordingly. Let $M(P,U)$ be the set of all perfect matchings in $G(P,U)$. We define the *matching dilation* of (P,O) as

$$\mathrm{MD}(P,O) = \min_{U \in \mathcal{L}} \ \min_{M \in M(P,U)} \ \max_{e \in M} \ w(e).$$

Since each particle has to move to some position in U for some $U \in \mathcal{L}$ to solve the coating problem, we have $\mathrm{OPT} \geq \mathrm{MD}(P,O)$. The search for the matching that minimizes the maximum edge cost for a given $U \in \mathcal{L}$ can be realized efficiently by reducing it to a flow problem using edges up to a maximum cost of c and performing binary search on c to find the minimal c such that a perfect matching exists. We note that our lower bound is not tight. This is due to the fact that it only respects the distances that particles have to move but ignores the congestion that may arise, i.e., in certain instances the distances to the object might be very small, but all particles may have to traverse one "chokepoint" and thus they block each other.

We implemented the Universal Coating algorithm in the amoebot simulator (see [20] for videos). The actual shape of a valid object O is irrelevant to our algorithm because particles only need to know where their immediate neighbors in the border around the object are relative to themselves, and this can be determined independently of the shape of the border. Thus for simplicity each simulation is initialized with a hexagon of object particles. We then initialize the particle system as idle particles attached randomly around the hexagon's perimeter. Parameters of the implementation are the radius of the hexagon that has to be coated and the number of (initially idle) particles that will run the algorithm. Each experimental trial randomly generates a new initial configuration of the system. Figure 2(a) shows the number of rounds needed to complete the coating. The radius of the static hexagon corresponds to the series label and the number of active particles is varied on the x-axis. The number of rounds are averaged over 20 trials of a constant size system. The confidence intervals plotted represent 95 %. These results show that in practice the number of rounds increases linearly with the number of particles in the system. This make sense because leader election takes a constant number of rounds for a given radius, so the increase comes only from the layering step. Figure 2(b) shows the ratios of the results from Fig. 2(a) compared to the lower bound defined above. The results shown in Fig. 2(b) show that the average competitive ratio of our algorithm approximated using the average ratio of the actual runtimes of the algorithm over the respective lower bounds (as defined above) may exhibit closer to logarithmic behaviors. Figure 2(c) shows the round duration of the algorithm as the radius of the static hexagon is varied. The runtime of the algorithm appears to increase linearly with both number of active particles and the size of the object being coated. There is also increased variability for systems with larger radii.

References

1. Derakhshandeh, Z., Gmyr, R., Strothmann, T., Bazzi, R., Richa, A.W., Scheideler, C.: Leader election and shape formation with self-organizing programmable matter. In: Phillips, A., Yin, P. (eds.) DNA 2015. LNCS, vol. 9211, pp. 117–132. Springer, Heidelberg (2015)
2. Derakhshandeh, Z., Dolev, S., Gmyr, R., Richa, A.W., Scheideler, C., Strothmann, T.: Brief announcement: amoebot - a new model for programmable matter. In: ACM SPAA, pp. 220–222 (2014)
3. Derakhshandeh, Z., Gmyr, R., Richa, A.W., Scheideler, C., Strothmann, T.: Universal coating for programmable matter. Theoretical Computer Science, http://dx.doi.org/10.1016/j.tcs.2016.02.039 (2016). Also appears as arXiv:1601.01008
4. Lynch, N.A.: Distributed Algorithms. Morgan Kaufmann, San Francisco (1996)
5. Doty, D.: Theory of algorithmic self-assembly. Commun. ACM **55**(12), 78–88 (2012)
6. Patitz, M.J.: An introduction to tile-based self-assembly and a survey of recent results. Nat. Comput. **13**(2), 195–224 (2014)
7. Woods, D.: Intrinsic universality and the computational power of self-assembly. In: Machines, Computations and Universality, pp. 16–22 (2013)
8. Angluin, D., Aspnes, J., Diamadi, Z., Fischer, M.J., Peralta, R.: Computation in networks of passively mobile finite-state sensors. Distrib. Comput. **18**(4), 235–253 (2006)
9. Bonifaci, V., Mehlhorn, K., Varma, G.: Physarum can compute shortest paths. In: ACM SODA, pp. 233–240 (2012)
10. Li, K., Thomas, K., Torres, C., Rossi, L., Shen, C.-C.: Slime mold inspired path formation protocol for wireless sensor networks. In: Dorigo, M., et al. (eds.) ANTS 2010. LNCS, vol. 6234, pp. 299–311. Springer, Heidelberg (2010)
11. Wilson, S., Pavlic, T., Kumar, G., Buffin, A., Pratt, S.C., Berman, S.: Design of ant-inspired stochastic control policies for collective transport by robotic swarms. Swarm Intell. **8**(4), 303–327 (2014)
12. Brambilla, M., Ferrante, E., Birattari, M., Dorigo, M.: Swarm robotics: a review from the swarm engineering perspective. Swarm Intell. **7**(1), 1–41 (2013)
13. Kumar, G.P., Berman, S.: Statistical analysis of stochastic multi-robot boundary coverage. In: ICRA, pp. 74–81 (2014)
14. Pavlic, T., Wilson, S., Kumar, G., Berman, S.: An enzyme-inspired approach to stochastic allocation of robotic swarms around boundaries. In: ISRR, pp. 16–19 (2013)

15. Blázovics, L., Csorba, K., Forstner, B., Charaf, H.: Target tracking and surrounding with swarm robots. In: ECBS, pp. 135–141 (2012)
16. Blázovics, L., Lukovszki, T., Forstner, B.: Target surrounding solution for swarm robots. In: Szabó, R., Vidács, A. (eds.) EUNICE 2012. LNCS, vol. 7479, pp. 251–262. Springer, Heidelberg (2012)
17. Michail, O., Spirakis, P.G.: Simple and efficient local codes for distributed stable network construction. In: ACM PODC, pp. 76–85 (2014)
18. Derakhshandeh, Z., Gmyr, R., Porter, A., Richa, A.W., Scheideler, C., Strothmann, T.: On the runtime of universal coating for programmable matter (2016). arXiv:1606.03642
19. Derakhshandeh, Z., Gmyr, R., Richa, A.W., Scheideler, C., Strothmann, T.: An algorithmic framework for shape formation problems in self-organizing particle systems. In: NANOCOM, pp. 21:1–21:2 (2015)
20. amoebot.cs.upb.de

Time Complexity of Computation and Construction in the Chemical Reaction Network-Controlled Tile Assembly Model

Nicholas Schiefer and Erik Winfree[✉]

California Institute of Technology, Pasadena, CA 91125, USA
winfree@caltech.edu

Abstract. In isolation, chemical reaction networks and tile-based self-assembly are well-studied models of chemical computation. Previously, we introduced the chemical reaction network-controlled tile assembly model (CRN-TAM), in which a stochastic chemical reaction network can act as a non-local control and signalling system for tile-based assembly, and showed that the CRN-TAM can perform several tasks related to the simulation of Turing machines and construction of algorithmic shapes with lower space or program complexity than in either of its parent models. Here, we introduce a kinetic variant of the CRN-TAM and investigate the time complexity of computation and construction. We analyze the time complexity of decision problems in the CRN-TAM, and show that decidable languages can be decided as efficiently by CRN-TAM programs as by Turing machines. We also give a lower bound for the space-time complexity of CRN-TAM computation that rules out efficient parallel stack machines. We provide efficient parallel implementations of non-deterministic computations, showing among other things that CRN-TAM programs can decide languages in $\mathsf{NTIME}(f(n)) \cap \mathsf{coNTIME}(f(n))$ in $\mathcal{O}(f(n) + n + \log c)$ time with $1 - \exp(-c)$ probability, using volume exponential in n. Lastly, we provide basic mechanisms for parallel computations that share information and illustrate the limits of parallel computation in the CRN-TAM.

1 Introduction

Biological organisms create remarkably sophisticated structures through the interplay of genetically-encoded chemical reactions and molecular self-assembly. DNA nanotechnology is beginning to explore the analogous potential of information-based chemistry by developing programmable circuitry using DNA strand displacement cascades [7, 20, 21, 25, 35], programmable self-assembly using DNA tile systems [3, 12, 22, 29, 31], as well as systems that combine both dynamic circuitry and self-assembly processes [33, 34]. Whereas there are well-developed theoretical models for dynamic chemical circuits [5, 18, 26, 27] and tile self-assembly [9, 23, 28] within which questions about the algorithmic power and efficiency of such systems can be posed and answered, the interplay of chemical reaction and self-assembly processes has received relatively little theoretical attention.

© Springer International Publishing Switzerland 2016
Y. Rondelez and D. Woods (Eds.): DNA 2016, LNCS 9818, pp. 165–182, 2016.
DOI: 10.1007/978-3-319-43994-5_11

It has been understood for decades that enzymes acting on information-bearing polymers can in principle perform efficient Turing-universal computation [4,6,13] and recently plausible molecular implementations using DNA nanotechnology have been proposed [14,19], but these studies do not exploit the full power of two- or three-dimensional self-assembly, nor do they explicitly concern themselves with how chemical reaction network computation can direct the construction of complex structures. We recently introduced a theoretical model [24], called the Chemical Reaction Network-Controlled Tile Assembly Model (CRN-TAM) that integrates the formal chemical reaction network (CRN) model [26] and the abstract tile assembly model (aTAM) [23]. We proved that in this model, the interplay between chemical reactions and self-assembly enables more efficient computation (in terms of space used) and more efficient construction (in terms of program size and shape scale) than either of the previous models alone. However, some of our constructions—devised to facilitate the proofs—were obviously inefficient in terms of time; they failed to exploit the inherent parallelism of molecular systems. Here, our goal is to determine whether integrating chemical reaction dynamics and tile self-assembly also provides as dramatic an advantage in terms of speed, for both computation and construction.

Using our previous definition of the structure and semantics of the CRN-TAM, we formulate a kinetic model based on the Gillespie dynamics of stochastic chemical reaction networks. Through a natural notion of CRN-TAM composition, we introduce the notion of an efficient encoding of an input and a fixed CRN-TAM decider that decides strings in a language, leading to the notion of copy-tolerant CRN-TAM deciders that can be readily composed and operated in parallel without interference. With reference to our previous stack machine construction, we show that there are space-efficient CRN-TAM deciders that use as much volume as a Turing machine would use space. We also give a lower bound showing that there are no space-efficient and copy-tolerant CRN-TAM deciders.

Our main result and most significant technical contribution is Theorem 5, which shows that there are copy-tolerant tile-based CRN-TAM deciders, which we demonstrate using a sentinel process like the one used by Adleman et al. [1] for analyzing the time complexity of assembling squares in a variant of the aTAM. To show how these copy-tolerant CRN-TAM deciders can be used for efficient parallel computation, we give a randomized CRN-TAM program that generates all 2^k strings of length k efficiently and with exponentially small error. We then combine these results to show that CRN-TAM programs can efficiently simulate non-deterministic computations in parallel, allowing them to (probabilistically) decide problems in $\mathsf{NTIME}(f(n)) \cap \mathsf{coNTIME}(f(n))$ in nearly $\mathcal{O}(f(n))$ time. Lastly, we show how the copy-tolerant CRN-TAM decider can be extended to allow limited state sharing between computations executing in parallel and discuss some likely limitations of parallelism in the CRN-TAM.

2   Semantics and Kinetics of the CRN-TAM

We begin by briefly reviewing the fundamental definition of the CRN-TAM, which was defined in detail by Schiefer and Winfree [24].

Definition 1. *A* tile *is an oriented square with a bond on each of the north, east, south, and west sides. Each of these* bonds *is a distinct tuple (ℓ, s), with a label ℓ drawn from some alphabet and non-negative integer strength s. Formally, a tile is a four-tuple $(\boxed{t}) = (N, E, S, W)$ of bonds for the north, east, south, and west sides, respectively. In this paper, tiles are denoted by symbols surrounded by boxes.*

As in the abstract tile assembly model, tiles can aggregate to form larger structures. These structures are held together by the bonds on the edges of the tiles; to join onto an assembly, every tile must be bound with at least a minimum binding strength, or *temperature*:

Definition 2. *An* assembly *is an aggregation of adjacent tiles; formally, an assembly composed of tiles from a set T is a function $A : \mathbb{Z}^2 \to (T \cup \{\varepsilon\})$ that assigns to each side of the 2D lattice a tile from T. If $A(x, y) = \varepsilon$, the site is empty or unoccupied. A function A is a* valid *assembly if and only if:*

- *The occupied sites of the assembly form a connected set.*
- *The origin $(0, 0)$ is occupied by a tile $A(0, 0) \in T$; this tile is called the* seed *of the assembly.*
- *The total binding strength of the tile at each non-empty site is at least the temperature τ.*

Assemblies are denoted by symbols surrounded by double boxes. For example, $\boxed{t}$ is a tile, but $\boxed{\boxed{A}}$ is an assembly. The (infinite) set of all valid assemblies using tiles from T is denoted M_T. When clear from context, $\boxed{\boxed{X}}$ is an assembly containing only the tile $\boxed{x}$ as its seed.

In the CRN-TAM, the notion of a tile and assembly are essentially equivalent to those in models derived from the abstract tile assembly model [2,17,23,28]. In models that include only tiles, a tile set T completely specifies the structural form of a tile-based program.

Definition 3. *A* CRN-TAM *program is a tuple (S, T, R, τ, I) consisting of:*

- *A finite set S of identified* signal *species. We also use a notational "empty" species ε and let $S_\varepsilon = S \cup \{\varepsilon\}$.*
- *A finite set T of tuples $\left(\boxed{t}, t^*\right)$ pairing tiles $\boxed{t}$ and their* removal *signals t^*, where $t^* \in S_\varepsilon$. The same tile may appear at most once in T, i.e. it cannot have two different removal signals.*
- *A finite set R of reactions, each of which is one of the following types:*
  - *CRN reactions $A + B \xrightarrow{k} C + D$, for signals $A, B, C, D \in S_\varepsilon$.*
  - *Tile deletion reactions $A + \boxed{t} \xrightarrow{k} C + D$, for signals $A, C, D \in S_\varepsilon$.*

- *Tile creation reactions $A + B \xrightarrow{k} \boxed{t} + C$ or $A + B \xrightarrow{k} \boxed{t} + \boxed{t'}$, for signals $A, B, C \in S_\varepsilon$ and tiles $\boxed{t}$ and $\boxed{t'}$.*
- *Tile relabelling reactions $A + \boxed{t} \xrightarrow{k} B + \boxed{t'}$ for signals $A, B \in S_\varepsilon$ and tiles $\boxed{t}$ and $\boxed{t'}$.*
- *Tile activation reactions $A + \boxed{x} \xrightarrow{k} \boxed{\boxed{x}} + x^*$ where $A \in S_\varepsilon$ and $\left(\boxed{x}, x^*\right) \in T$.*
- *Tile deactivation reactions $\boxed{\boxed{x}} + x^* \xrightarrow{k} A + \boxed{x}$ where $A \in S_\varepsilon$ and $\left(\boxed{x}, x^*\right) \in T$.*

In all of these reactions, a reactant or product ε indicates that the reactant or product does not exists; for example, a reaction $A + \varepsilon \xrightarrow{k} \varepsilon + D$ is just $A \xrightarrow{k} D$. The reaction rate constant k must be specified as a rational number.

- *The temperature $\tau \in \mathbb{N}$, which is the minimum binding strength.*
- *The initial state I, a multiset of tiles and signals that are initially present. Often, we will use I as a function $I : (S \cup T) \to \mathbb{N}$ indicating the number of a particular element in the multiset. An initial state does not contain any assemblies.*

Structurally, the signal species in S are analogous to the species of a stochastic chemical reaction network (sCRN) and the tiles in T are analogous to the tile set that defines a program in the abstract tile assembly model (aTAM). Signal species and unbound tiles float in a well-mixed vessel, interacting analogously to the species in a stochastic chemical reaction network.

Occasionally, it will be useful to *combine* several CRN-TAM programs.

Definition 4. *The combination of two CRN-TAM programs $P = (S, T, R, \tau, I)$ and $P' = (S', T', R', \tau, I')$ is $P \oplus P' = (S \cup S', T \cup T', R \cup R', \tau, I \cup I')$, so long as tiles are consistent, i.e., $\left(\boxed{t}, x\right) \in T$ and $\left(\boxed{t}, y\right) \in T'$ implies $x = y$. Note that since I and I' are multisets, duplicates are repeated in the union.*

As in the aTAM, tiles bind together to form assemblies provided that they attach with a total binding strength that is at least the temperature τ. Along with the reaction specified in R, there are implicit *addition and removal reactions* in the CRN-TAM that are similar to the corresponding reactions in other tile assembly models, but with a slight change in character; extended rationale is given in our previous paper introducing the CRN-TAM [24].

Definition 5. *Let $\left(\boxed{t}, t^*\right) \in T$, and let $\boxed{\boxed{A}}$ and $\boxed{\boxed{B}}$ be assemblies that differ by exactly $\boxed{t}$ in some location other than $(0,0)$. A tile addition reaction is a reaction*

$$\boxed{\boxed{A}} + \boxed{t} \xrightarrow{1} \boxed{\boxed{B}} + t^*$$

Notice that since $\boxed{\boxed{B}}$ is valid, $\boxed{t}$ is attached with total strength at least τ. The corresponding removal reaction

$$\boxed{\boxed{B}} + t^* \xrightarrow{1} \boxed{\boxed{A}} + \boxed{t}$$

may occur only when $\boxed{t}$ is bound by exactly strength τ, and the removal signal is not ε. Note that the seed tile is privileged and cannot be removed from the assembly; it may be deactivated only after all other tiles have been removed.

The signals, free tiles, and assemblies in the reaction vessel completely specify the state of a CRN-TAM program at any time:

Definition 6. *A state L of a CRN-TAM program P is a multiset of signals, tiles, and assemblies. As with the initial state I, we use the notation $L(x) : (S \cup T \cup M_T) \to \mathbb{Z}^+$ to refer to the current count of x in L.*

Frequently, we will refer to *the program state* or *current state*, which is simply the state that reflects the current contents of the reaction vessel.

Definition 7. *A reaction is* possible *for a state L if its rate constant is nonzero and for every one of its reactants α, $L(\alpha) > 0$. The possible reactions $\mathrm{Pos}(L)$ of a state L include all of the possible reactions in R and all of the possible tile addition and removal reactions. Note that $\mathrm{Pos}(L)$ is always finite.*

The possible reactions induce a graph that describes the possible transitions between different states.

Definition 8. *The reaction graph $G(P)$ of a CRN-TAM program $P = (S, T, R, \tau, I)$ is a directed graph with a vertex for each of the (infinitely numerous) states of P and a directed edge from L to all states in $\mathrm{Pos}(L)$ for all states L. The reachable reaction graph is a subgraph of $G(P)$ with only vertices that are descendants of the initial state I. Where it is unambiguous or unimportant, we may refer to the reachable reaction graph as simply the reaction graph.*

Note that reachable reaction graphs are by definition connected, but may not be strongly connected. As reactions occur, the program state will change from one state to the other in a manner that is governed by the reactions' propensities: in this paper, we are especially interested in CRN-TAM programs that eventually reach a state with no possible reactions, where they will remain forever.

Definition 9. *A state of a CRN-TAM program is a* termination state *if it has no possible reactions. Equivalently, the termination vertex has no out edges in the reaction graph. A CRN-TAM program P* stops *if it reaches a termination state with probability one and the set of reachable states is finite.*

As we will see in Definition 11, we cannot consider the temporal dynamics of a CRN-TAM program without knowing the volume in which the program operates. In this paper, we will always use a default volume, dependent on the program, that ensures that any execution path will at all times have a finite density, in the following sense:

Definition 10. *An atomic chemical species is a signal or a tile, whether free or bound to an assembly. The* mass *of a state is the total number of atomic chemical species present in all signals, tiles, and assemblies. The* volume *required by a CRN-TAM program P is the maximum mass present for any state in the reachable state graph of P.*

By this definition, if the reachable reaction graph is infinite, then it must have a state whose mass exceeds any given bound, and thus violates the finite density constraint for some possible execution path. Our choice of volume cannot handle such systems, and in this paper we restrict our attention only to CRN-TAM programs with finite reachable reaction graphs. In the CRN-TAM, as in stochastic chemical reaction networks [26], the volume is fixed at the beginning and does not change over time.

Thus far, we have specified only the possible reactions associated with a state and not the dynamics of the system. The program state evolves according to stochastic Gillespie dynamics, where reactions occur at a rate proportional to their current *propensity* [10, 11].

Definition 11. *Let $P = (S, T, R, \tau, I)$ be a CRN-TAM program in state L that uses volume V. The* propensity *of a reaction R with rate constant k is given by:*

– $\rho(R) = kL(R_1)$ *if R is a unimolecular reaction with reactant R_1.*
– $\rho(R) = kL(R_1)L(R_2)/V$ *if R is a bimolecular reaction with two distinct reactants R_1 and R_2.*
– $\rho(R) = kL(R_1)(L(R_1) - 1)/V$ *if R is a bimolecular reaction with identical reactants.*

Note that a reaction is possible if and only if its propensity is nonzero. With reference to Definition 11, we can finally define the kinetics of a CRN-TAM program.

Definition 12. *Let $L(t)$ be a random variable-value for the current state of a CRN-TAM program P at a time $t \in [0, \infty)$. The state $L(t)$ evolves over time as a continuous time Markov chain on the space of possible CRN-TAM states with (deterministic) initial state $L(0)$. For two distinct program states A and B, the transition rate between them is given by the propensity of the reaction in P that converts state A into state B, 0 if there is no such reaction, or the sum of propensities if there is more than one (e.g. $X \xrightarrow{1} Y$ and $C + X \xrightarrow{1} C + Y$).*

For each of stochastic CRNs and the abstract tile assembly model, there is a natural notion of the "size" of a molecular program; in sCRNs, this is the number of reactions, while in the aTAM it is the number of tiles. We can similarly define a measure of program complexity for the CRN-TAM.

Definition 13. *The* complexity *of an initial state $I : (S \cup T) \to \mathbb{N}$ is*

$$|I| = \sum_{z \in (S \cup T)} \log_2(I(z) + 1)$$

This definition is natural since it is the number of bits needed to specify a general initial state I, up to small constant multiplicative and additive factors. We similarly define the size of a set of reactions such that if all reactions have unit rate constants, it is just the count of reactions, but otherwise it scales as the information needed to specify the rates:

Definition 14. *The complexity of a set of reactions R, where $r \in R$ is written as $A_r + B_r \xrightarrow{k_r} C_r + D_r$ and $k_r = \frac{n_r}{d_r}$ as an irreducible fraction, is*

$$|R| = \sum_{r \in R} \log_2(n_r \times d_r + 1)$$

The complexity of a CRN-TAM program is the sum of the complexities of its components.

Definition 15. *Let $P = (S, T, R, \tau, I)$ be a CRN-TAM program. The complexity of P with respect to temperature τ is*

$$K_{CT}^\tau(P) = |S| + |T| + |R| + |I|$$
$$= |S| + |T| + \sum_{r \in R} \log_2(n_r \times d_r + 1) + \sum_{z \in (S \cup T)} \log_2(I(z) + 1)$$

Each term is related to the number of bits required to specify the corresponding part of the program, up to logarithmic factors. Like sCRNs but unlike the aTAM, we allow nontrivial initial state as a convenience; our previous work showed that for any CRN-TAM program P, there is a CRN-TAM program P' with no initial state and program complexity $K_{CT}^\tau(P') = \Theta(K_{CT}^\tau(P))$ that simulates it [24, Theorem 4].

3   Efficient Computation

We are principally concerned with using CRN-TAM programs to perform efficient computation. As in much of theoretical computer science, we deal primarily with decision problems, and therefore formulate a model of computation that solves them. As defined so far, CRN-TAM programs cannot "compute" in the sense that a Turing machine or even a circuit can; a program is fixed and has no notion of input. Furthermore, most of the efficient and natural encodings of fixed strings in the CRN-TAM involve assemblies, while a CRN-TAM program does not have any assemblies in its initial state. To resolve this, we begin by describing a natural way of encoding fixed input strings as CRN-TAM programs, and then demonstrate how those input strings can be combined with CRN-TAM *deciders* to provide a model of computation.

Definition 16. *Consider a CRN-TAM program P and a string x over an alphabet Σ. Let $T_\Sigma : T \to \Sigma$ be a partial function from the tiles of P to the alphabet, assigning some of the tiles in T to represent symbols in the alphabet. We say that P encodes x if it constructs a $1 \times (|x| + 1)$ rectangular assembly A with the following properties:*

- A begins with a designated "start tile" $\boxed{t_{\text{start}}}$.
- Let the kth tile after the start tile be $\boxed{t_k}$. Then $T_\Sigma\left(\boxed{t_k}\right) = x_k$, the kth symbol in the string.

Of course, a string of length n can always be encoded by a CRN-TAM program with $\Theta(n)$ complexity, by using a distinct tile type for each symbol in the string. Previously, Adleman et al. [1] and Soloveichik and Winfree [28] showed that strings of length n can be encoded[1] in aTAM tile sets with $\mathcal{O}(n/\log n)$ distinct tile types at temperature $\tau = 2$. We can show that CRN-TAM programs with sublinear complexity can encode n-symbol strings. In [24] it was shown that:

Theorem 1. *For any string x over a constant-size alphabet Σ and temperature $\tau \geq 1$, there is a CRN-TAM program $P = (S, T, R, \tau, I)$ that encodes x and has complexity $K_{\text{CT}}^\tau(P) = \mathcal{O}(n/\log n)$.*

Moreover, this bound is tight. We therefore have a natural definition of input for computation with the CRN-TAM:

Definition 17. *An* input encoding *of a string x (over an constant alphabet) is a CRN-TAM that encodes x and has complexity $K_{\text{CT}}^\tau(P) = \mathcal{O}(|x|/\log|x|)$ at any temperature $\tau \geq 1$.*

Our CRN-TAM computers should be independent of the input and should solve an appropriate decision problem when combined with an input encoding. For the rest of the paper, we will use a default encoding $E_\Sigma(x)$ for alphabet Σ and input string x, which we will simply refer to as $E(x)$ where the alphabet is clear from context.

Definition 18. *Consider a language $\mathcal{L}$ that is decidable by a Turing machine M. Consider a fixed CRN-TAM program D with two identified signals Q_{accept} and Q_{reject}. We say that D is a* CRN-TAM decider *for $\mathcal{L}$ with respect to the default input encoding E if, for every input x the combined program $D \oplus E(x)$:*

- *Produces Q_{accept} and then stops immediately if and only if M accepts x.*
- *Produces Q_{reject} and then stops immediately if and only if M rejects x.*

For convenience, we will say that D accepts *x if the first case holds and* rejects *x if the second case holds.*

For any execution of a CRN-TAM decider acting on an input, there is a time t^* for which $L(t) = L(t^*)$ for all $t \geq t^*$. Using the dynamics described in Definition 12, we can use this to define the amount of time that a computation in the CRN-TAM takes.

---

[1] By necessity, a different notion of "encoding" must be used in the aTAM, since building even a $1 \times n$ rectangle requires $\Theta(n)$ tile types [2]. However, the notion used in the aTAM is analogous to our notion of encoding.

Definition 19. *Consider a CRN-TAM decider D for language $\mathcal{L}$, and a string x. The* time *that $D + E(x)$ takes to decide x is the random variable T^* for the minimum time t^* so that $L(t) = L(t^*)$ for all $t \geq t^*$. We call this the* stopping time *of the CRN-TAM program. As we are usually interested in asymptotically characterizing the worst-case time that a decider takes to decide all inputs of a given length, we define the random variable $T(n) = \max_{x \in \Sigma^n} T^*(x)$.*

The aTAM is an inherently scalable model of molecular computation, in the sense that we may consider an arbitrary number of assemblies growing in parallel and executing independent computations. Because there is no mechanism for inter-assembly interaction and the supply of tiles is fixed (and implicitly infinite), each assembly is a universe unto itself; from the perspective of an aTAM programmer, it does not matter if a reaction vessel has a single assembly or a million. Like stochastic CRNs, however, the CRN-TAM has a shared global state. As a result, a CRN-TAM program does not necessarily scale for parallel execution: combining just two functioning CRN-TAM programs may not produce a functioning CRN-TAM program. Nonetheless, scalability is a desirable property from a theoretical standpoint, since it might allow parallel computations, and from a practical standpoint, where any molecular program is unlikely to have an isolated reaction vessel. As a seemingly minimal base case, a CRN-TAM program that can scale for parallel execution ought to still work correctly when multiple copies of the same program act on the same input. We are therefore especially interested in CRN-TAM deciders that are copy-tolerant, in the following formal sense that is closely related to the similarly-named notions for CRNs [8,15]:

Definition 20. *A CRN-TAM decider D for language $\mathcal{L}$ is* copy-tolerant *if for any $k \geq 2$, $D'_k = \bigoplus_{i=1}^{k} D$ is also a CRN-TAM decider for $\mathcal{L}$.*

Intuitively, a copy-tolerant CRN-TAM decider is one that supports running multiple instances of the decider in the same reaction vessel simultaneously and still reports the answer accurately. As we will see, many convenient CRN-TAM deciders are not copy-tolerant, and there appear to be substantive lower bounds on the volume required by copy-tolerant CRN-TAM deciders.

4   Space-Efficient Deciders

Definition 21. *A* site *(i, j) containing tile $\boxed{x}$ is* immediately dependent *on site (i', j') containing tile $\boxed{x'}$ if it shares a bond with $\boxed{x'}$ and was added to the assembly after $\boxed{x'}$. The site (i, j) is* dependent *on (i', j') if it shares a bond with a tile that is dependent or immediately dependent on $\boxed{x'}$.*

The recursive definition of dependency induces a directed, acyclic graph of dependencies, where edges go from tiles to the tiles they are immediately dependent on. In such a graph, a tile's descendants are all the tiles it is dependent on, and a tile's ancestors are all those tiles that depend on it.

Definition 22. *Consider an assembly A. The dependency graph of A is a directed graph with the occupied sites of A as vertices and a directed edge from each site to the sites on which it is immediately dependent.*

As an immediate corollary of Definition 22, the dependency graph is acyclic and rooted at the seed of the assembly, as proved in [24]. Dependency is a critical concept for attempting to disassemble an assembly; to disassemble, we are forced to remove only leaves of the dependency DAG, performing the entire disassembly in "dependency-reversed" order.

Theorem 2. *A tile cannot be removed until all of its ancestors in the dependency graph have been removed. That is, a tile cannot be removed until all tiles at sites dependent on it have been removed.*

Theorem 2 has both reassuring and limiting consequences. On one hand, it ensures that tiles cannot be ripped out from the middle of the assembly when their removal signals are present. On the other, it implies that we can never create "temporary scaffolding" for our CRN-TAM constructions: a CRN-TAM program may never build parts of an assembly that are dependent on scaffolded parts that are meant to be removed later.

If a language $\mathcal{L}$ can be decided on a Turing machine using space $s(n)$, we might hope that there is a CRN-TAM decider for L that uses only $\Theta(s(n))$ volume. Indeed, the existence of such deciders is an immediate corollary of the stack machine construction from [24], itself a modification of the stack polymer construction by Qian et al. [19].

Theorem 3. *Given a language $\mathcal{L}$ decided by a Turing machine in space $s(n)$, there is a CRN-TAM decider for $\mathcal{L}$ that uses $\Theta(s(n))$ volume.*

The stack machine construction relied critically on storing the state of the stack machine in the global CRN and is therefore not copy-tolerant. We might hope to construct a CRN-TAM decider that is similarly space efficient, using asymptotically as much volume as a Turing machine uses space on its tape, but that also remains copy-tolerant. Unfortunately, this turns out to be impossible.

Theorem 4. *Consider a language $\mathcal{L}$ decidable by a Turing machine that requires $s(n)$ space and $t(n)$ time. Every copy-tolerant CRN-TAM decider for $\mathcal{L}$ uses volume $\Omega(t(n))$.*

Proof. Omitted due to lack of space.   □

In general, there are many functions where $t(n) \in \omega(s(n))$, i.e. that require much more time to compute than space, and so there are languages for which space efficient, copy-tolerant CRN-TAM deciders do not exist.

5   Time Complexity of Tile Computations

Having established that stack machine-based CRN-TAM deciders experience a slowdown proportional to the volume and that copy-tolerant deciders require $\Omega(t(n))$ volume, we safely abandon attempts to build copy-tolerant stack machine-type deciders. If we have temperature at least 2, where cooperative binding is possible, we can also perform Turing-universal computation using tiles alone [23,30,32]. This approach proves to be very fruitful, giving us a key theorem:

Theorem 5. *For every language $\mathcal{L}$ decidable in time $t(n)$ and using space $s(n)$ on a Turing machine, there is a copy-tolerant CRN-TAM decider D for $\mathcal{L}$ with expected time complexity $\Theta(t(n))$ and volume complexity $\Theta(t(n)s(n))$.*

Proving Theorem 5 is somewhat technical and involves some careful stochastic analysis. The key problem is that CRN-TAM programs have only a finite supply of each tile type present at any given moment. In the aTAM and many of its derivatives, this is not a problem, since an infinite supply of tiles with fixed ratios is generally assumed. The issue is further complicated by the finite density constraint and the desire for *fast* computation; having all the necessary tiles present from the beginning would raise the volume prohibitively and make computation needlessly slow. To resolve this in the CRN-TAM, we use a simple mechanism for regenerating consumed tiles; although the mechanism is intuitive, the asynchronicity of efficient computation as well as fluctuations in the tile concentrations make it harder to analyze. Our proof that this mechanism leads to efficient computation—as our intuitions tell us—involves analyzing a *sentinel process* like the one introduced by Adleman et al. [1] in a variant of the aTAM, where we artificially constrain the dynamics to create a stochastic process that is easy to analyze but still stochastically dominates the actual CRN-TAM dynamics. We analyze the dynamics of the sentinel process as a phase-type distribution on CRN-TAM states. Lastly, we apply some careful combinatorics and the Chernoff bound for exponentially distributed random variables to show Theorem 5.

At a conceptual level, our efficient and copy-tolerant CRN-TAM decider will use a constant-sized Turing-universal aTAM tile set to perform the computation proper, by the well-known method of computation histories (Fig. 1). To ensure that the (bimolecular) tile addition reactions proceed quickly, the decider will use a tree-structured counter to efficiently generate an $\Theta(V)$ concentration of each of those (constant in number) tiles. Lastly, the tile removal signal will be consumed after its release and used to produce a new tile, which will replace the consumed tile in the "pool" of available tiles. Each of these conceptual points will be shown as a separate lemma, along with a number of technical lemmas used for the analysis of the sentinel process.

To begin, we adapt a classic result from the aTAM [30], to show that there are effective tile-only CRN-TAM deciders which make only minimal use of the CRN.

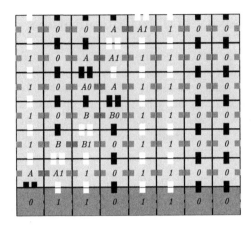

Fig. 1. An illustration of the assembly built by the tile-based CRN-TAM decider TuringTiles, in this case implementing a Turing machine with two states (A and B) operating on the binary alphabet $\Sigma = \{0, 1\}$ and the rule that the read head goes left whenever it sees a 0 and goes right whenever it sees a 1, flipping each bit as it goes. The input string (01101100) is shown at the bottom.

Lemma 1. *For any decidable language, there is a CRN-TAM program*

$$\mathsf{TuringTiles} = (\{Q_{\mathrm{accept}}, Q_{\mathrm{reject}}\}, T, 2, \varnothing)$$

where Q_{accept} is the removal signal for a tile t_{accept} that indicates acceptance and Q_{reject} is the removal signal for a tile t_{reject} that indicates rejection, consisting of only tiles and the necessary signals, that is a CRN-TAM decider when combined with an initial state containing a large enough supply of tiles.

Proof. Omitted due to lack of space.   □

In the CRN-TAM, it is critically important to have a high concentration of tiles for addition since tile addition reactions are bimolecular. By definition, assemblies have unit concentration, so we must aim to have $\Theta(V)$ tiles of a given type in the reaction vessel of volume V in order to have constant expected time for tile addition. For the case considered by Theorem 5, however, at the beginning neither the input encoding nor the CRN-TAM decider contains enough tiles; thankfully, they can be generated efficiently.

Lemma 2. *Given a species (signal or tile) A, there is a CRN-TAM program $\mathsf{TreeCounter}_n(A)$ that stops with 2^n copies of A in $\mathcal{O}(n)$ expected time, with program complexity $K_{\mathrm{CT}}^1(\mathsf{TreeCounter}_n(A)) \in \mathcal{O}(n)$.*

Proof (sketch). For every $0 \le i \le n$, introduce the signal S_i and the reaction $S_i \to S_{i+1} + S_{i+1}$ unless $i = n$, with S_0 as the sole species present in the initial state. Notice that the total number of S_i produced is precisely twice the number of S_{i-1} that were produced, so by induction a total of 2^n instances of S_n are produced by this program. By adding the reaction $S_n \to A$, we are guaranteed to eventually stop with precisely 2^n copies of A.

Through methods similar to those that will be used in the proof of Theorem 5, we can show that $\mathsf{TreeCounter}_n(A)$ stops in $\mathcal{O}(n)$ expected time. The key insight is that every reaction in the tree counter is unimolecular and so can always proceed at rate $\Omega(1)$.   □

Lastly, there is a simple mechanism for regenerating tiles, which we name for convenience.

Definition 23. *For a tile* $(\boxed{t}, t^*)$ *in a CRN-TAM program, we define the program* $\mathsf{ReplaceTile}_\tau(\boxed{t}, t^*) = (S, T, R, \tau, I) = \left(\{t^*\}, \{(\boxed{t}, t^*)\}, \{t^* \to \boxed{t}\}, \tau, \varnothing\right).$

Combining Lemmas 1 and 2, we can efficiently generate a $\Theta(V)$ concentration for all of the tiles in the (constant-sized) tile set for computation in only $\Theta(\log V)$ time; so long as the concentration remains that high, tile addition reactions will happen in constant expected time and each tile addition will release a removal signal t^* that will be converted back into an active tile in constant expected time. We now have the preliminaries necessary to state our construction of the CRN-TAM decider in Theorem 5:

Proof (construction for Theorem 5). Consider a CRN-TAM program $D = (S, T, R, 2, I)$ consisting of the combination of $\mathsf{TuringTiles}$ and, for all tiles $(\boxed{t}, t^*) \in T$, both $\mathsf{TreeCounter}_{\lceil \log V \rceil}(\boxed{t})$ and $\mathsf{ReplaceTile}_2(\boxed{t}, t^*)$. Since the tree counter will produce $\Theta(V)$ tiles of each tile type right from the beginning, the combined program is a CRN-TAM decider by Lemma 1. Lastly, notice that D is copy-tolerant, since the entirety of the computation happens on a single assembly.   □

The analysis of the expected time for D to decide L, including several technical lemmas, will appear in the full paper.

6   Combinatorial Assembly Production and Nondeterministic Parallelism

The copy-tolerant CRN-TAM decider in Theorem 5 allows us to perform several threads of computation in parallel, given sufficient volume. In general, it is simple to make a copy of the input for each thread efficiently, using a tree-style copier like the one used in Lemma 2. However, identical inputs are not useful; any deterministic tile set such as the tile-based CRN-TAM decider would produce precisely the same computation in all of the parallel threads. To remedy this, we might hope to generate identifiers for each of the k threads as they are produced during the input copying process. To this end, we could perform a *combinatorial assembly* task, like generating all 2^k binary strings of length k in a serial fashion as was done in [24], but this would take time exponential in k. If we are willing to allow some small chance of error, a very simple CRN-TAM program can assemble these strings in $\Theta(k)$ time.

Theorem 6. *For any positive integer* c, *there is a* CRN-TAM *program* CombinGenerate(n) *with complexity* $K^1_{CT}($CombinGenerate(n)$) \in \Theta(cn)$ *that stops having constructed all* 2^n *assemblies, each of size* $1 \times (n + 1)$, *encoding all binary strings of length* n, *in time* $\Theta(\log(cn2^n))$ *with probability* $1 - e^{-c}$.

Proof. Omitted due to lack of space.   □

We would of course prefer to be able to do combinatorial string assembly deterministically and efficiently. Intuitively, this seems extremely difficult, for the following reason. Suppose that we have assembled all but one of the strings; if we are operating with some kind of parallelism, how can we know which string has not yet assembled without some kind of exponential communication problem? While we have not been able to prove anything beyond a few special cases, we suspect efficient, deterministic combinatorial assembly is not possible.

Conjecture 1. There is no CRN-TAM program that constructs all 2^n binary strings of length n and runs in $\mathcal{O}(\text{poly}(n))$ time.

As mentioned earlier, this combinatorial assembly can be used for generating various *seeds* for parallel computations, so that different parallel threads can operate differently. A simple application of this concept is implementing parallel non-determinism, where each seed acts as a string of binary guesses for a nondeterministic Turing machine.

Theorem 7. *For any language* $\mathcal{L} \in$ NTIME($f(n)$)∩coNTIME($f(n)$) *and positive integer* c, *there is a* CRN-TAM *decider for* $\mathcal{L}$ *that decides it in* $\Theta(f(n)+n+\log c)$ *expected time, with probability at least* $1 - e^{-c}$.

Proof (sketch). First, observe that an input encoding can be converted into one that generates 2^f copies of the input by replacing each production of every tile $\boxed{t}$ with an instance of TreeCounter$_f(\boxed{t})$ instead. Per our analysis of the tree counter, this takes only $\mathcal{O}(f)$ time, so we can generate $cn2^n$ copies of the input in only $\mathcal{O}(n + \log c)$ time. Observe that a nondeterministic Turing machine running in $f(n)$ time can use at most $f(n)$ bits of nondeterminism, so we need only generate all $2^{f(n)}$ such strings. Consider a modification of CombinGenerate($f(n)$) from above where the $f(n)$th bit has a glue that matches the seed of an input assembly, so that the input assemblies will grow at the end of each combinatorially assembled seed. By modifying the tree counters that raise the initial tile concentration, that we can modify the CRN-TAM decider from Theorem 5 to generate $cf(n)2^{f(n)}$ times as many tiles, so that the concentration of each tile is still $\Theta(V)$. Lastly, we modify the reject signal so that it is simply ε. So long as every string of nondeterministic choices is generated by CombinGenerate($f(n)$), the system will produce the accept signal if and only if the input x is in $\mathcal{L}$. We can perform a similar operation on the Turing machine that decides the complement of $\mathcal{L}$, except that we change its accept signal to the reject signal. Combining these two programs, we obtain a CRN-TAM decider for $\mathcal{L}$ that runs in $\Theta(f(n)+n+\log c)$ expected time, and succeeds whenever CombinGenerate($f(n)$) produces all $2^{f(n)}$ strings of length $f(n)$.   □

More concretely, we find that there are languages that are decidable in poly-nomial time by CRN-TAM programs that are not decidable in polynomial time by Turing machines under standard complexity theoretic assumptions. Notice however that exponential *volume* might still be required; while we might evalu-ate all nondeterministic branches in parallel, each still needs space to operate.

Corollary 1. *Any language in* NP ∩ coNP *has a CRN-TAM decider that runs in* $\Theta(\mathsf{cpoly}(n))$ *and succeeds with probability at least* $1 - e^{-c}$.

On its own, Theorem 7 demonstrates the computational power of the CRN-TAM. A careful reader will note that from the perspective of an actual laboratory experiment, the construction does not offer much beyond the capabilities of tile-only systems based on the aTAM. Although the theoretical formulation of the aTAM does not permit multiple assemblies, all experimental realizations thus far have many, many assemblies forming in the same reaction vessel [3,22]. Furthermore, the same tile set used for the combinatorial seed production step works just as well in the aTAM. From this point of view, the advantages offered by the CRN-TAM are nice, but not fundamentally different. Unlike the aTAM tile set, the CRN-TAM program can detect when an answer has been computed and exponentially amplify a signal indicating that. As we will see, the biggest advantage of the CRN-TAM is that is allows interaction between the assemblies while it is computing, allowing more powerful forms of parallel computation.

7   Towards Parallel Computation with Shared State and Open Questions

Even within the framework we have already described, a form of elementary shared state can be implemented with only a slight modification. Consider adding a special state tile with a distinguished removal signal ϕ and a reac-tion $\phi \to \boxed{\phi}$ that converts it into a tile that can bind to another state tile. If one assembly moves into the appropriate state, it can release ϕ and, by hav-ing $\boxed{\phi}$ attach onto another assembly, share some constant number of bits of information with another assembly. Furthermore, this signal can be amplified very quickly using a tree counter, allowing it to "turn the test tube red" and inform, with arbitrarily high probability, every other assembly that some assem-bly reached state ϕ. Rudimentary branch-and-bound, an algorithmic technique that has proved immensely useful for gaining dramatic speedups in optimization problems with only exponential time algorithms, can be implemented with this kind of rudimentary mechanism. In effect, we can use the CRN-TAM to simulate the types of parallel steps introduced by Lipton [16] for classical DNA computing that consisted of a series of laboratory operations on test tubes of DNA data, thus replacing a non-autonomous molecular computation by an autonomous one.

The parallel computational power of the CRN-TAM is far from limitless, however. Although we have been unable to show any concrete lower bounds, it seems very difficult to implement many general forms of parallel computation—such as languages decided by uniform circuits in time proportional to their depth, for example—because the well-mixed CRN is a difficult medium for passing information. In particular, sending more than constant-sized messages between different assemblies seems very difficult, since we must rely on chance to have the message arrive at its destination, and every other recipient must somehow recognize that the message is intended for another assembly.

Our work here has established a convenient kinetic model for the CRN-TAM and analyzed the time complexity of basic computational primitives, but many important questions remain open. Although we have shown the lower bound in Theorem 4, there is a gap between our lower bound (that copy-tolerant CRN-TAM deciders require $\Omega(t(n))$ volume) and our best construction, which requires $\mathcal{O}(t(n)s(n))$ space. The question of whether there exist copy-tolerant CRN-TAM deciders that require $o(t(n)s(n))$ volume remains open. Most questions related to efficient parallel computation in the CRN-TAM also remain open, and we have only considered the most basic ways of implementing it.

Acknowledgements. We acknowledge financial support from National Science Foundation grant CCF-1317694 and the Soli Deo Gloria Summer Undergraduate Research Fellowship at the California Institute of Technology. We also thank Dave Doty and Damien Woods for their insights.

References

1. Adleman, L., Cheng, Q., Goel, A., Huang, M.D.: Running time and program size for self-assembled squares. In: Proceedings of the 33rd Annual ACM Symposium on Theory of Computing, STOC 2001, pp. 740–748 (2001)
2. Aggarwal, G., Cheng, Q., Goldwasser, M.H., Kao, M.Y., de Espanes, P.M., Schweller, R.T.: Complexities for generalized models of self-assembly. SIAM J. Comput. **34**, 1493–1515 (2005)
3. Barish, R.D., Schulman, R., Rothemund, P.W.K., Winfree, E.: An information-bearing seed for nucleating algorithmic self-assembly. Proc. Natl. Acad. Sci. **106**, 6054–6059 (2009)
4. Bennett, C.H.: The thermodynamics of computation - a review. Int. J. Theoret. Phys. **21**, 905–940 (1982)
5. Cardelli, L.: Two-domain DNA strand displacement. Math. Struct. Comput. Sci. **23**, 247–271 (2013)
6. Cardelli, L., Zavattaro, G.: On the computational power of biochemistry. In: Horimoto, K., Regensburger, G., Rosenkranz, M., Yoshida, H. (eds.) AB 2008. LNCS, vol. 5147, pp. 65–80. Springer, Heidelberg (2008)
7. Chen, Y.J., Dalchau, N., Srinivas, N., Cardelli, L., Soloveichik, D., Seelig, G.: Programmable chemical controllers made from DNA. Nat. Nanotechnol. **8**, 755–762 (2013)
8. Condon, A., Kirkpatrick, B., Maňuch, J.: Reachability bounds for chemical reaction networks and strand displacement systems. Nat. Comput. **13**, 499–516 (2014)
9. Doty, D.: Theory of algorithmic self-assembly. Commun. ACM **55**, 78–88 (2012)
10. Gillespie, D.T.: Exact stochastic simulation of coupled chemical reactions. J. Phys. Chem. **81**, 2340–2361 (1977)
11. Gillespie, D.T.: Stochastic simulation of chemical kinetics. Annu. Rev. Phys. Chem. **58**, 35–55 (2007)

12. Ke, Y., Ong, L.L., Shih, W.M., Yin, P.: Three-dimensional structures self-assembled from DNA bricks. Science **338**, 1177–1183 (2012)
13. Kurtz, S., Mahaney, S., Royer, J., Simon, J.: Biological computing. In: Complexity Theory Retrospective II, pp. 179–195 (1997)
14. Lakin, M.R., Phillips, A.: Modelling, simulating and verifying turing-powerful strand displacement systems. In: Cardelli, L., Shih, W. (eds.) DNA 17 2011. LNCS, vol. 6937, pp. 130–144. Springer, Heidelberg (2011)
15. Lakin, M.R., Stefanovic, D., Phillips, A.: Modular verification of chemical reaction network encodings via serializability analysis. Theoret. Comput. Sci. **632**, 21–42 (2016)
16. Lipton, R.J.: DNA computations can have global memory. In: International Conference on Computer Design: VLSI in Computers and Processor, pp. 344–347 (1996)
17. Patitz, M.J.: An introduction to tile-based self-assembly and a survey of recent results. Nat. Comput. **13**, 195–224 (2013)
18. Phillips, A., Cardelli, L.: A programming language for composable DNA circuits. J. R. Soc. Interface **6**, S419–S436 (2009)
19. Qian, L., Soloveichik, D., Winfree, E.: Efficient turing-universal computation with DNA polymers. In: Sakakibara, Y., Mi, Y. (eds.) DNA 16 2010. LNCS, vol. 6518, pp. 123–140. Springer, Heidelberg (2011)
20. Qian, L., Winfree, E.: Scaling up digital circuit computation with DNA strand displacement cascades. Science **332**, 1196–1201 (2011)
21. Qian, L., Winfree, E., Bruck, J.: Neural network computation with DNA strand displacement cascades. Nature **475**, 368–372 (2011)
22. Rothemund, P.W.K., Papadakis, N., Winfree, E.: Algorithmic self-assembly of DNA Sierpinski triangles. PLoS Biol. **2**, e424 (2004)
23. Rothemund, P.W.K., Winfree, E.: The program-size complexity of self-assembled squares. In: Proceedings of the 32nd Annual ACM Symposium on Theory of Computing, STOC 2000, pp. 459–468 (2000)
24. Schiefer, N., Winfree, E.: Universal computation and optimal construction in the chemical reaction network-controlled tile assembly model. In: Phillips, A., Yin, P. (eds.) DNA 2015. LNCS, vol. 9211, pp. 34–54. Springer, Heidelberg (2015)
25. Seelig, G., Soloveichik, D., Zhang, D.Y., Winfree, E.: Enzyme-free nucleic acid logic circuits. Science **314**, 1585–1588 (2006)
26. Soloveichik, D., Cook, M., Winfree, E., Bruck, J.: Computation with finite stochastic chemical reaction networks. Nat. Comput. **7**, 615–633 (2008)
27. Soloveichik, D., Seelig, G., Winfree, E.: DNA as a universal substrate for chemical kinetics. Proc. Natl. Acad. Sci. **107**, 5393–5398 (2010)
28. Soloveichik, D., Winfree, E.: Complexity of self-assembled shapes. SIAM J. Comput. **36**, 1544–1569 (2007)
29. Wei, B., Dai, M., Yin, P.: Complex shapes self-assembled from single-stranded DNA tiles. Nature **485**, 623–626 (2012)
30. Winfree, E.: On the computational power of DNA annealing and ligation. In: DNA Computers. DIMACS Series in Discrete Mathematics and Computer Science, vol. 27, pp. 199–221. American Mathematical Society (1996)
31. Winfree, E., Liu, F., Wenzler, L.A., Seeman, N.C.: Design and self-assembly of two-dimensional DNA crystals. Nature **394**, 539–544 (1998)
32. Winfree, E., Yang, X., Seeman, N.C.: Universal computation via self-assembly of DNA: some theory and experiments. In: DNA Based Computers II. DIMACS Series in Discrete Mathematics and Computer Science, vol. 44, pp. 191–213. American Mathematical Society (1999)

33. Yin, P., Choi, H.M.T., Calvert, C.R., Pierce, N.A.: Programming biomolecular self-assembly pathways. Nature **451**, 318–322 (2008)
34. Zhang, D.Y., Hariadi, R.F., Choi, H.M.T., Winfree, E.: Integrating DNA strand-displacement circuitry with DNA tile self-assembly. Nat. Commun. **4**, Article no. 1965 (2013)
35. Zhang, D.Y., Turberfield, A.J., Yurke, B., Winfree, E.: Engineering entropy-driven reactions and networks catalyzed by DNA. Science **318**, 1121–1125 (2007)

Author Index

Printed in the United States
By Bookmasters